Just Cause: Marine Operations in Panama 1988-1990

by
Lieutenant Colonel Nicholas E. Reynolds
U. S. Marine Corps Reserve

HISTORY AND MUSEUMS DIVISION
HEADQUARTERS, U.S. MARINE CORPS
WASHINGTON, D.C.
1996

PCN 190 003134 00

For sale by the U.S. Government Printing Office
Superintendent of Documents, Mail Stop: SSOP, Washington, DC 20402-9328
ISBN 0-16-048729-3

Foreword

The history of Marines in Panama from 1988 to 1990, the years bracketing Operation Just Cause, does not involve a great many Marines, and Marines made up only a small percentage of the forces that went into battle during the operation itself in December 1989. But for a period of more than two years, the Marines in Panama were literally in foxholes on the front lines. For the individual Marine, going to Panama meant going in harm's way. Although most planners thought of Panama in terms of low intensity conflict, the personal experience for many Marines was very intense, and tested their courage, endurance, and professionalism.

One measure of the intensity of Just Cause lies in the fact that the contribution to the success of the operation by Marine Forces Panama was out of proportion to its size—that is, greater by far than for most other units of comparable size, measured in terms of prisoners captured or objectives seized. At the same time, the experience in Panama yielded some interesting lessons about low intensity conflict, including lessons in mobility and patrolling that an earlier generation of Marines who served in Central America would have understood.

To research and preserve the story of Marines in Panama, the History and Museums Division deployed Benis M. Frank in the spring of 1991. Mr. Frank, who was then head of the Oral History Section and is now Chief Historian in the division, walked the ground and interviewed Marines who participated in the operation. After his return to the United States, he conducted further interviews and supervised the collection of Marine Forces Panama records.

Once Mr. Frank had fulfilled the collection phase, a member of Mobilization Training Unit (Historical) DC-7, which supports the History and Museums Division, Lieutenant Colonel Nicholas E. Reynolds, USMCR, assumed responsibility for the project, and completed the history. Lieutenant Colonel Reynolds joined the Marine Corps in 1975 after receiving a doctorate in history from Oxford University, where he wrote a book on the German Army. Following active duty with 3d Battalion, 2d Marines, he held a variety of reserve billets, including that of company commander at the Basic School during and after the Persian Gulf War.

In the pursuit of accuracy and objectivity, the History and Museums Division welcomes comments on this publication from interested individuals and activities.

MICHAEL F. MONIGAN
Colonel, U.S. Marine Corps
Acting Director of Marine Corps History and Museums

Preface

This story is about the Marines who served in Panama around the time of Operation Just Cause. Since the Marine forces comprised only a fraction of the troops in Panama, their contribution has been overlooked in some other histories. This is especially true of the Marines who served in Panama before and after the operation itself. Nevertheless, they faced, and met, a very real set of challenges of their own, and wrote one of the first chapters in the Marine Corps' history of operations other than war since the fall of the Berlin Wall. The Marines who patrolled the jungles of the Canal Zone in the period before Just Cause, in what was neither peace nor war, broke new ground. So did the young officers and NCOs who, for all intents and purposes, took over the reins of municipal government after the operation.

Since Benis M. Frank, the Chief Historian in the History and Museums Division, had done a thorough job of interviewing the participants and collecting documents, both in the field and from sources in Washington, my job was relatively easy. I was able to rely almost exclusively on what is now the Marine Corps Historical Center's collection on Panama, including transcripts of Mr. Frank's interviews, as well as a variety of plans, operations, and reports. I also consulted a number of command chronologies and secondary sources, and corresponded with a few of the Panama Marines.

Throughout, I was able to rely on the support and advice of many members of the staff of the History and Museums Division. Both the Director, Brigadier General Edwin H. Simmons, and Mr. Frank himself were kind enough to read the manuscript and give me the feedback I needed. Similarly, members of other sections provided valuable assistance: Danny J. Crawford, head of the Reference Section; Joyce M. Conyers of the Archives Section; Evelyn A. Englander, the librarian; John T. Dyer, the art curator; and Charles R. Smith of the History Writing Unit, who edited the final manuscript. The former Division Deputy Director, Colonel William J. Davis, gave new meaning to the term "brother officer." I am grateful to him and all the other Marines who helped bring the project to completion. Nevertheless, I alone am responsible for the opinions expressed, and any errors that may appear in the text.

NICHOLAS E. REYNOLDS
Lieutenant Colonel
U. S. Marine Corps Reserve

Table of Contents

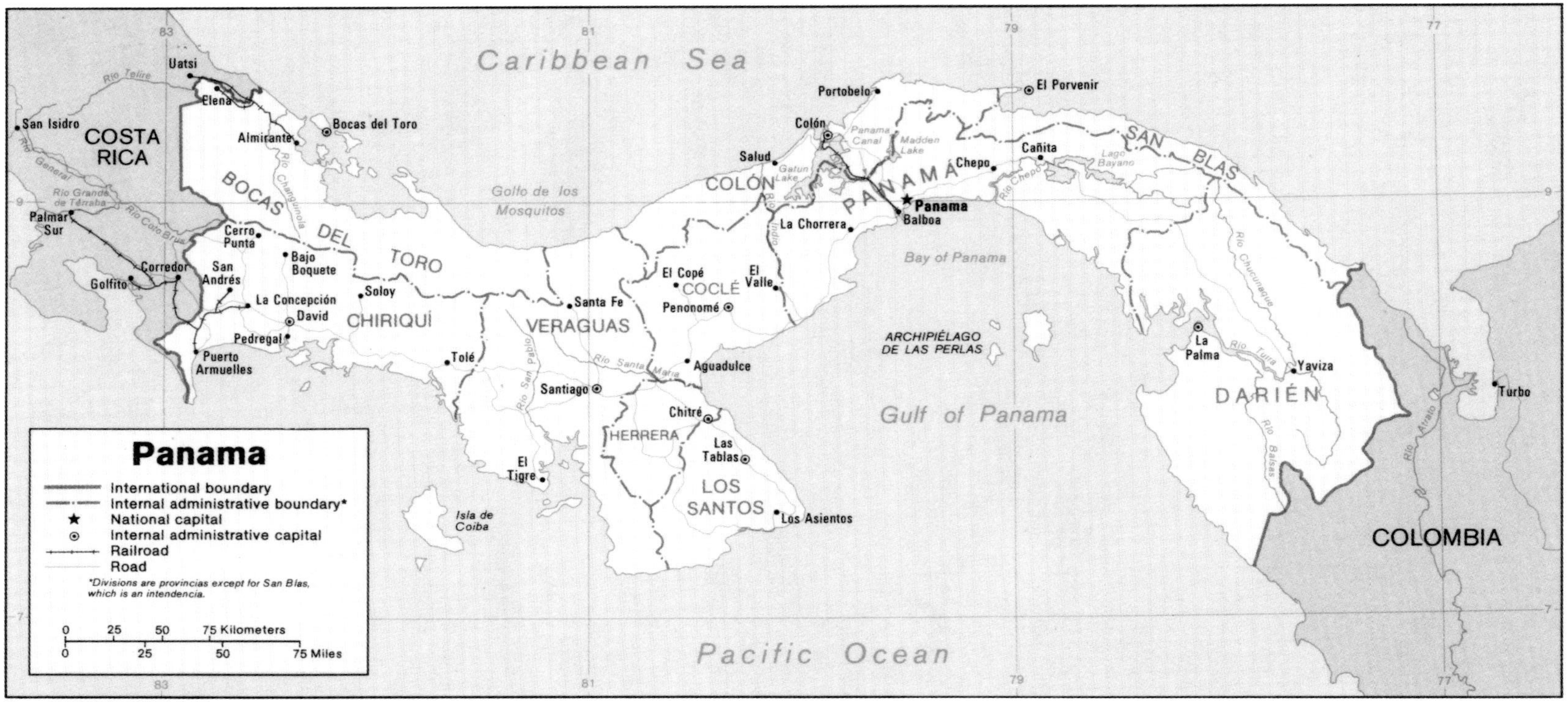

Base 504908 (540285) 8-81

Just Cause: Marine Operations in Panama 1988-1990

First Reinforcements — First Firefight — Recovery and Consolidation — Neither War Nor Peace — The Evolution of Training — The New Year — The October Coup and its Aftermath — Setting the Stage for Just Cause — Combat — La Chorrera — PDF Prisoners — More New Responsibilities — Christmas Eve — Christmas Day and After — Marine Reinforcements — Operation Promote Liberty

---

Marines first came to Panama in 1856 to protect fortune hunters on their way to California across the Isthmus, and returned a number of times for brief periods, typically to restore law and order, between 1856 and 1900. When Panama seceded from Colombia in 1903, in large part to make it possible for the United States to acquire the rights for a canal under favorable terms, the Marine Corps landed forces under Major John A. Lejeune to guarantee Panamanian independence. Lejeune was one of the first to realize that the Marines had come to Panama to stay, and, in 1904, his troops set up a semi-permanent barracks near Panama City. Until 1911, their primary mission was to safeguard the canal that was under construction.[1]

By October 1923, a permanent Marine Barracks in Panama was established at the U.S. Naval Submarine Base, Coco Solo. Over the following decades, Marine forces in Panama went through periods of expansion and contraction, reaching a peak of 36 officers, 3 warrant officers, and 1,571 enlisted men in February 1945. They also went through a number of redesignations and relocations. In 1943, Marines in Panama were consolidated under Marine Barracks, Fifteenth Naval District, which was renamed Marine Barracks, Rodman, Canal Zone, in 1976. In August 1987, the official title was changed to Marine Corps Security Force (MCSF) Company, Panama. Nevertheless, the standing complement of approximately five officers and 125 Marines remained at U.S. Naval Station Panama Canal, also known as Rodman Naval Station, which was located on the western shore of the Canal near the Pacific Ocean exit.[2]

The title suggested the mission. By now, the primary mission of the Marines in Panama was to protect local naval installations, while that of U.S. Army units in the Canal Zone was to protect the Canal itself. In 1988, Army units operated under the umbrella of U.S. Army South (USArSo), which was in turn a component of Southern Command (SouthCom). Both commands had their headquarters in Panama.

Over the years since 1903, relations with Panama, which lay on either side of the American-controlled Canal Zone, were not always amiable. Due to the location of the Canal and its control by a foreign power, Panamanian interests clashed with those of the United States from time to time. However, the signing of the Panama Canal Treaty in 1977, which provided for complete Panamanian control of the Canal by the year 2000, appeared to guarantee future cooperation between the two countries.

The prospects for good relations, however, dimmed when General Omar Torrijos, the Panamanian leader who negotiated and signed the treaty, died in a plane crash in 1981. His death opened the way for a troublesome successor, General Manuel Antonio Noriega, who contrived to control the nation through figurehead politicians from his position as head of the country's armed forces. In 1983, Noriega became head of the National Guard, which he soon combined with the Air Force and Navy to create the Panamanian Defense Forces (PDF).

By 1988, the PDF was as much a social and political system as a military force. It was a conglomerate of police, military, and paramilitary organizations with a total strength in the neighborhood of 15,000. However, it could only field some 3,000 to 3,500 combat troops, largely trained and equipped by the U.S. Army. On the ground, the PDF had two battalions in each of the country's 13 military zones, in addition to 10 independent companies, a cavalry squadron, and a handful of special forces. The PDF's air arm had roughly 50 aircraft, while its navy could deploy 12 small vessels. Complementing Panamanian Defense Forces were the so-called "Dignity Battalions." Supposedly created to counter American aggression, the Dignity Battalions were little more than formations of laborers, many of whom were unemployed, with an overlay of military discipline. There were 14 such battalions, each of which could deploy some 200 to 250 members.

Although Noriega had a history of cooperating with American intelligence agencies, his involve-

Watercolor by Anthony F. Stadler

Referred to as the "Big House," the headquarters building of Marine Barracks, Panama, was built in 1965. Although headquartered at Rodman Naval Station, the barracks included a number of separate detachments scattered throughout the Canal Zone.

ment in the drug trade eventually made it impossible for the U.S. Government to continue to work with him. Relations between the two countries worsened markedly when, in February 1988, two Florida federal grand juries indicted him on charges of racketeering and trafficking in narcotics. Due in part to the indictments, Panamanian President Eric Arturo Delvalle attempted to depose Noriega, but without success. Instead Noriega engineered Delvalle's dismissal. The result was further civil disorder and both implicit and explicit threats to American lives and property, including incursions into U.S. Navy installations.

First Reinforcements

The United States reacted to the heightened tensions first by updating contingency plans for Panama and then by sending reinforcements to the Canal Zone. The plans included provisions for a Marine Expeditionary Brigade to deploy to Panama as part of Operation Elaborate Maze. But the Pentagon was not ready for Elaborate Maze, and the first contingent of Marine reinforcements was a great deal smaller, intended only to strengthen Marine security forces. The initial contingent was a platoon from Fleet Anti-Terrorist Security Team (FAST) Company, Marine Corps Security Force Battalion, Atlantic, based in Norfolk, Virginia. A kind of military SWAT team, the FAST platoon was trained in close-quarter battle techniques. Arriving in Panama on 14 March, the platoon was placed under the operational control of the MCSF.

Major Eddie A. Keith, the security force's commanding officer, decided to use the platoon to protect the Naval Station's Arraijan Tank Farm (ATF), where fuel was stored in 37 underground tanks for use by all American forces in Panama. The tank farm covered approximately two square kilometers of rolling grassland, apparently designed to resemble a golf course from the air, but was surrounded by dense jungle, which provided excellent avenues of approach both to the storage tanks and to Howard Air Force Base to the south. In the

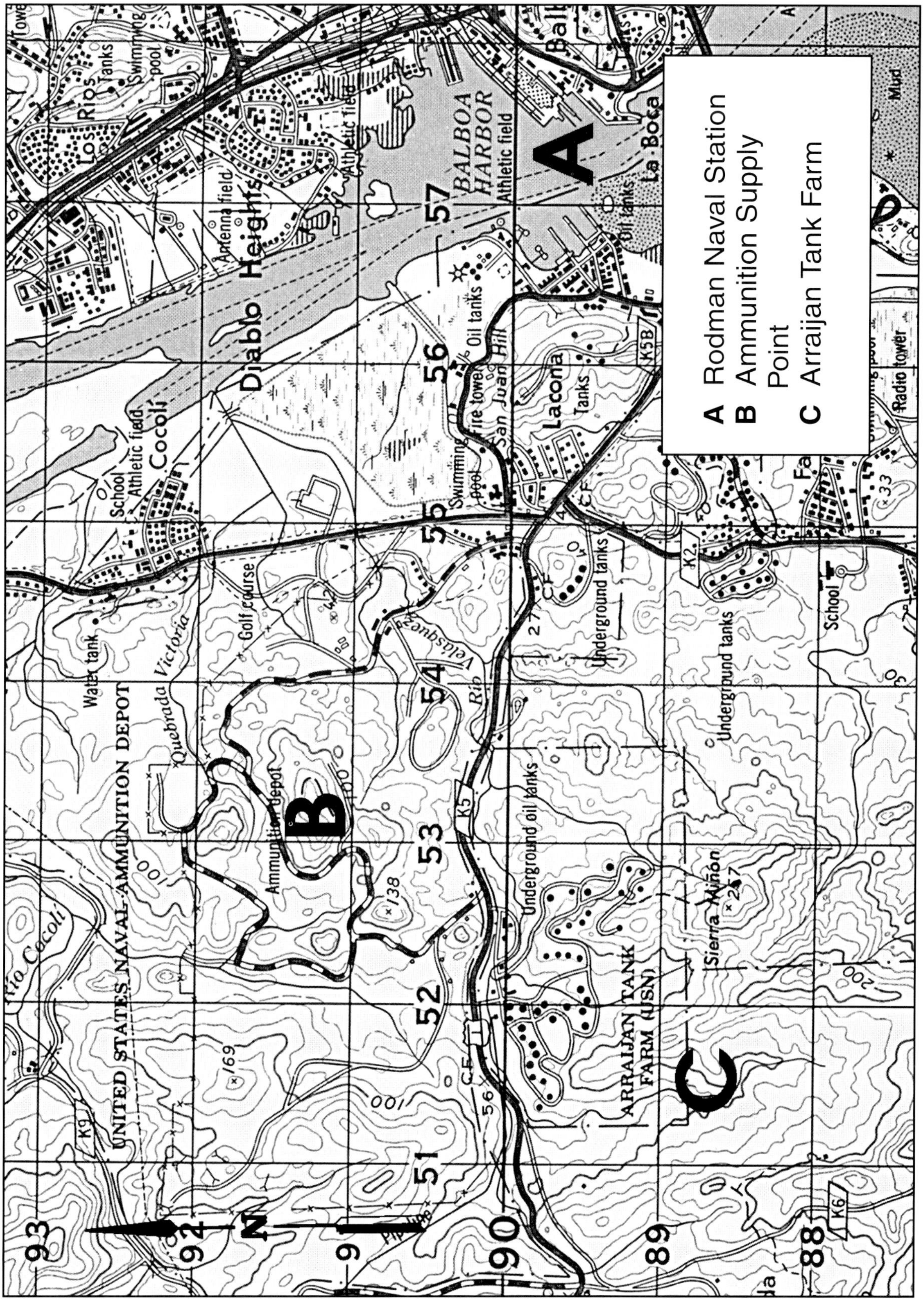
A Rodman Naval Station
B Ammunition Supply Point
C Arraijan Tank Farm
UNITED STATES NAVAL AMMUNITION DEPOT
Ammunition Depot
ARRAIJAN TANK FARM (USN)
BALBOA HARBOR
Diablo Heights
Cocoli
La Boca
Los Rios
Golf course
Underground oil tanks
Underground tanks
Sierra Miñon
Water tank
Radio tower
Mud

jungle, visibility was limited to a few feet, even by day, and movement was slow and exhausting. The rule of thumb was that a patrol could cover no more than 500 meters in one hour. The ATF was bounded on the north by the Naval Ammunition Depot, or ammunition supply point (ASP), which was fenced. But for the most part the tank farm was not fenced as it bordered the Pan-American, or Thatcher, Highway, the best high-speed avenue of approach in the country. Under the circumstances, it was nearly impossible for Marine security forces, with their limited resources, to guard the tank farm properly, let alone both the fuel and ammunition storage facilities.

Soon after its deployment to the Arraijan Tank Farm, the platoon reported the presence of intruders, usually at night. The Marines described the intruders as individuals wearing black camouflage uniforms, carrying weapons and night vision devices. Daily patrols found freshly dug fighting holes. When the security force reported its findings through the chain of command, the standard response was that Panamanian Defense Forces did not have night vision devices and that U.S. Army units had dug the holes during training exercises. However, the Marines conducted further research, and discovered that no Army units had trained in the vicinity in recent months, and that the Army had sold night vision devices to Panama during the past decade.[3]

While the FAST platoon patrolled the ATF, the political climate continued to worsen and the 6th Marine Expeditionary Brigade (MEB) staff at Camp Lejeune, North Carolina, prepared for possible deployment to Panama in accordance with contingency plan Elaborate Maze. Although Southern Command showed little, if any, enthusiasm for further Marine reinforcements, 6th MEB planners assumed an "all or nothing" approach. Should reinforcements be ordered, the brigade was prepared to deploy most of its strength, to include two infantry battalions, a reinforced helicopter

Marines patrol up a small stream in the jungle surrounding the Arraijan Tank Farm. The terrain provided intruders with good cover and avenues of approach both to the fuel storage facility and Howard Air Force Base.

Watercolor by Anthony F. Stadler

A panoramic view of Rodman Naval Station reveals its situation on the northern shore of Balboa Harbor, from where it controlled the Pacific entrance to the Canal.

squadron, and a detachment of OV-10 Bronco aircraft. On 31 March, Fleet Marine Force, Atlantic (FMFLant), ordered the MEB to deploy in accordance with a Joint Chiefs of Staff directive. The following day, an advance party of three brigade staff officers flew to Panama to coordinate the movement of forces.[4]

On 1 April, the Pentagon formally announced that 1,300 additional U.S. troops would be sent to Panama, including 300 Marines, once again simply to enhance security in view of the growing tension and unrest. Since the full strength of the brigade would not be needed, FMFLant, nevertheless, decided that the 300 Marines would deploy as a provisional Marine Air-Ground Task Force (MAGTF) built around a reinforced rifle company under the administrative control of 6th MEB. The idea was for the MAGTF to serve as cadre for the eventual deployment of the entire brigade. For that reason, the MEB supplied the command element, while 2d Marine Division provided a rifle company and 2d Force Service Support Group, a group of 40 Marines, to form Brigade Service Support Group 6 (BSSG-6) SouthCom's commander, General Frederick F. Woerner, USA, decided that the presence of a large number of Army and Air Force aircraft in Panama precluded the need for any Marine air assets. The MAGTF, therefore, was not a true "air-ground" task force, a fact which some Marine commanders, wholeheartedly committed to the MAGTF concept, regretted despite the fact that the Marines in Panama were more of a security force than a maneuver element.[5]

Marine reinforcements began flowing into Panama on 6 April, under the command of the brigade's chief of staff, Colonel William J. Conley, the senior officer present. Colonel Conley arrived in Panama to facilitate the movement of the brigade, not to command. However, since Southern Command did not see the need for additional Marine forces beyond the reinforced rifle company, and since the Pentagon decided to stop well short of executing Elaborate Maze, the MEB commander never deployed, and Colonel Conley assumed command.

The provisional MAGTF initially came under the

operational control of U.S. Navy South (USNavSo), and the rifle company, Company I, 3d Battalion, 4th Marines, began to operate in direct support of the commander of Rodman Naval Station, where the MAGTF planted its flag. Soon thereafter, an agreement was reached between Colonel Conley and Rear Admiral Jerry G. Gnecknow, USN, USNavSo commander, that he, Colonel Conley, would assume control of all Marine ground forces in the event of an emergency. The understanding formed the basis for the eventual establishment of Marine Forces Panama (MarForPM), whereby Colonel Conley assumed operational control of all Marine forces in country. Neither the agreement between Gnecknow and Conley nor the mission of Company I changed when, on 9 April, operational control of Marine Forces Panama shifted from U.S. Navy South to the newly established Joint Task Force Panama (JTF Panama).[6]

First Firefights

Company I was chosen to deploy to Panama in large part because it was available and because it was on air alert. The Company was well trained, having recently cycled through a combined arms exercise at Twentynine Palms, California; cold weather training at Bridgeport, California, and Fort McCoy, Wisconsin; and jungle training on Okinawa, as well as a Marine Corps Combat Readiness Evaluation. Most recently, its parent battalion had rehearsed for the Panama contingency, practicing missions such as defending a fuel farm or antenna site and dealing with intruders. Much of the training was on company commander's time, forcing commanders to rely on their own judgment and giving them the opportunity to mold their companies into cohesive, professional forces.[7]

Before deploying, the company was reinforced by an 81mm mortar section of two mortars, a surveillance and target acquisition (STA) platoon, a counter-intelligence team, and a squad of engineers. Arriving in country, the Marines made themselves as comfortable as possible in the Rodman gymnasium, while their commander, Captain Joseph P. Valore, conferred with Major Keith, the security force commander. Since it appeared that the company would deploy to the tank farm, Valore then conducted a reconnaissance of the area and debriefed members of the FAST platoon. On 7 April, Colonel Conley gave Captain Valore a rough operations order for the defense of the tank farm, as well as other nearby installations.[8]

Captain Valore used the information he obtained to analyze the situation and the mission. He realized that he would need to plan carefully, since his mission was ambitious for a reinforced rifle company, given the terrain and the threat. He concluded that the best way to cover the terrain and to minimize the threat to his Marines was to use patrolling techniques, as opposed to manning fixed positions. He knew that the "old" way, once described as "one man standing post in a little shack underneath a light surrounded by jungle," was a thing of the past.[9] Initially, he planned to use a system of patrol bases, whose locations would change at regular intervals. However, he soon decided that, rather than waste time and effort moving command posts and communications equipment, it would be more efficient to find and improve one set of good positions. Accordingly, Captain Valore divided the tank farm into two zones, and assigned each zone to one platoon as a tactical area of responsibility.[10]

Before deploying his Marines, Valore had to address a critical issue–weapons policy. Security force policy prohibited Marines from having a round in the chamber under routine conditions. Concerned for the safety of his Marines, and determined to demonstrate his trust in their judgment, Valore raised the issue with Colonel Conley. Conley agreed to permit Valore's Marines to patrol with a loaded magazine in their weapons and a round in the chamber. Valore's policy made sense. Under then-current rules of engagement, Marines were permitted to return fire, but if they did not have a magazine in the weapon and a round in the chamber, they might not live to return fire.[11]

The Marines' weapons policy was a highly charged issue. Although Valore focussed more on the immediate threat than on the past, a number of more senior Marine officers, including Lieutenant General Ernest T. Cook, Commanding General, FMFLant, remembered the 1983 tragedy at Marine Amphibious Unit Headquarters in Beirut, Lebanon. Then, a restrictive weapons policy was one of the contributing factors which made it impossible for sentries to stop the truck bomb which took the lives of 241 Marines, sailors, and soldiers. Determined not to get "stung" twice, Cook and his staff endorsed more aggressive and forward-leaning policies as that of Captain Valore, both in the

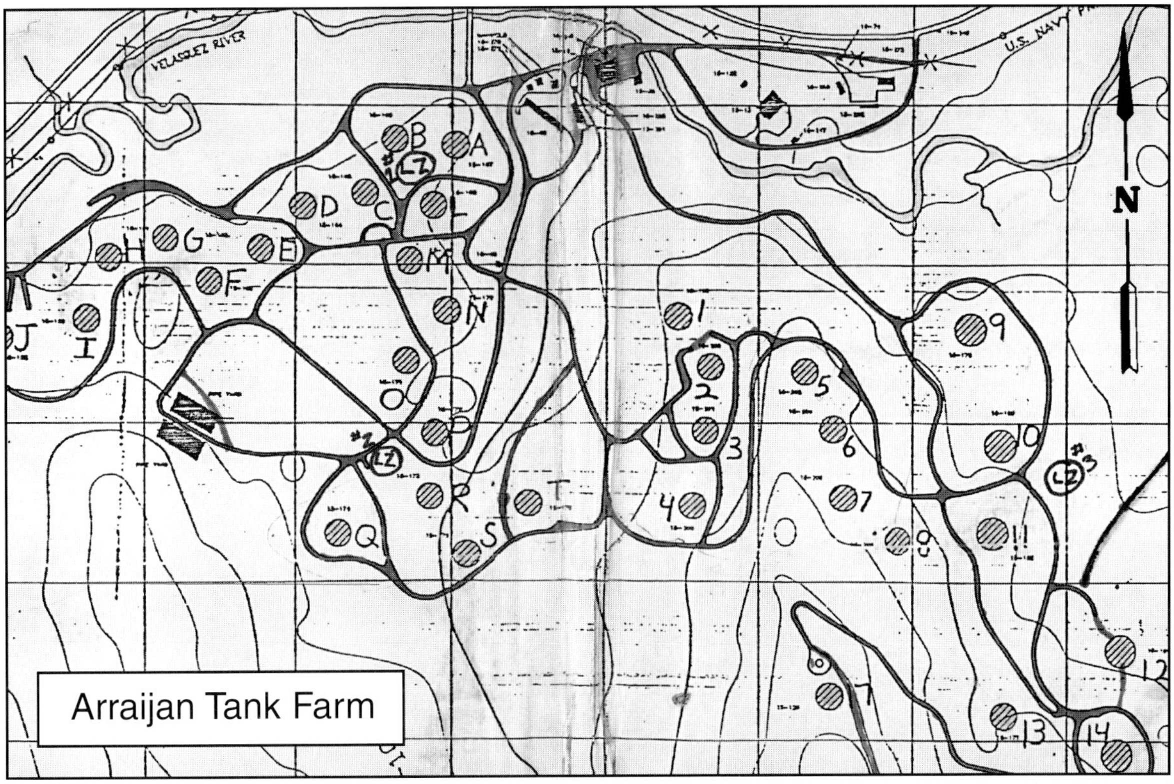

field and around Marine offices and housing on Rodman.[12]

Southern Command viewed the policy differently. General Woerner was under orders from Washington both to protect American interests and not to exacerbate the situation. It was a delicate and difficult balance. At the same time, long-time members of the U.S. military community in Panama argued that the Marines were over-reacting to the threat and that their aggressiveness could provoke a response in kind from the PDF. Marine commanders defended their policy by pointing out that they occupied exposed positions in the jungle, where they faced a threat from intruders which predated their weapons policy. The Marines, they pointed out, were responding to an old threat, not creating a new one.

In one form or another, the dispute continued throughout the existence of Marine Forces Panama. Whatever the merits on either side of the argument, it was understandable that there would be cultural differences between the Marines and the other Southern Command troops. Unlike most of their American counterparts in Panama, the Marines were combat troops on unaccompanied 90-day tours, except for Colonel Conley and principal members of his staff, who were on 180-day tours. As opposed to living in base housing and having the freedom of the city, the Marines worked long hours, usually under field conditions, for up to two weeks at a time. The two weeks of duty typically were followed by a two-day rest period, which might include some base liberty but never involved liberty off base, a precaution to prevent incidents between Panamanian Defense Forces and individual Marines.

Events soon transpired to test the assumptions of the Marines' weapons policy. On 9 and 10 April, shortly after the two platoons from Company I relieved the FAST platoon at the tank farm, unknown intruders began to probe their lines. On the night of 11 April, a squad-sized Marine patrol in the vicinity of "K" tank in the northeast sector of the farm made contact. The patrol split into two groups in an attempt to trap the intruders. When a flare misfired, igniting with a "pop" like a gunshot, one element opened fire with M-16 rifles, mistaking the flare for an intruder's weapon. Unfortunately, they fired in the direction of the other element, mortally wounding Corporal Ricardo M. Villahermosa, the patrol leader. Villahermosa was evacuated to Gorgas Army Community Hospital, where he died in the early morning hours of 12 April.[13]

The incident renewed the debate over the weapons policy among Marine and Army ground commanders. At a meeting on the 12th with the Joint Task Force commander, Major General Bernard Loeffke, USA, whose primary duty was commander of U.S. Army South, the only Marine present, Major Alfred F. Clarkson of the MEB staff, vigorously defended the Marine policy, citing guidance from the Commandant of the Marine Corps and adding his own view that it was morally wrong to send Marines in harm's way without fully loaded weapons. Colonel Conley fully supported Clarkson's position, and made certain that Marine commanders had the authority they needed to implement the right weapons policy for the conditions they faced.[14]

The average Marine rifleman, whose morale was already shaken by Villahermosa's death, had much the same worry–that he would now be ordered back into the jungle without a loaded weapon. Captain Valore put that worry to rest by visiting the line platoons and telling his Marines that the policy had not changed. As he had in the past, he made it clear that he had full confidence in their judgment and professionalism. Valore's actions did much to restore morale and re-energize the company.[15]

At dusk on the 12th, U.S. Army remote battlefield sensors on loan to the Marines detected movement by approximately 40 persons who appeared to be approaching the tank farm from the direction of the Pan-American Highway. The sensor activation was confirmed by the detachment of surveillance and target acquisition Marines from 3d Battalion, 4th Marines.* The detachment, in place at listening posts in the western sector of the ATF approximately 700 meters from the company command post, reported by radio that they had seen and heard intruders. A specially-equipped Air Force AC-130 Specter orbiting the area also confirmed the sightings.[16]

Captain Valore left his position in order to investigate the reports, and spotted approximately 12 intruders between the highway and the tank farm. At that point, he decided to begin consolidating his forces in the center of the ATF. Soon thereafter, his Marines received and returned fire, aiming down the line of tracers that came at them. The 13-man surveillance detachment, still in position to the west of the company, reported that the intruders were probing their positions, and Sergeant Michael A. Cooper, the noncommissioned officer in charge, immediately requested illumination. Eighty-one millimeter mortars near the company command post fired the mission. Putting the illumination to good use, Cooper reported that he could see that the intruders were armed and well-equipped, and that they moved like professional soldiers. When the intruders continued to advance despite the illumination rounds, he urgently called for the mortars to drop high-explosive rounds on pre-plotted targets near his position. Knowing that Cooper was a professional who would not overreact, Captain Valore authorized the fire missions, and the mortars fired 16 high-explosive rounds in addition to the 98 rounds of illumination they had already fired. He also authorized Cooper's Marines to return fire. Valore could hear the results, heavy firing, almost immediately.[17]

A few minutes later, around 1930, the company itself took fire from a ravine to its front. Valore directed company First Sergeant Alexander J. Nevgloski to return fire with a Mark 19 chain gun, which then fired a total of 222 40mm rounds. Within minutes, the fire from the Mark 19 along with the fire from the mortars effectively suppressed the threat, and firing died down.[18]

Sometime after 2200, General Loeffke arrived on scene, wearing civilian clothes and wanting to know what happened. After being briefed by Captain Valore, he ordered the company to cease fire and not to reengage unless fired upon. He also ordered the Marines to remain in place and to permit any intruders to leave the area. Loeffke said that he was in touch with the PDF, whose commanders had assured him that they did not have any troops in the area. Captain Moises Cortizo, an English-speaking PDF officer, stood next to General Loeffke, and restated the Panamanian position.[19]

After receiving General Loeffke's order, Valore pulled back Cooper's detachment from the western sector of the tank farm while the rest of his Marines held their positions. Using night vision devices they watched as a number of the intruders apparently received first aid, and saw others, either wounded or dead, being evacuated. Sensor activations confirmed their sightings. There was further

---

*The Army's remote battlefield sensor system apparently was operated by a detachment of Sensor Control and Management Platoon (SCAMP) Marines from the 2d Marine Division which deployed with Captain Valore's company.

corroboration. Security force Marines, who had manned a roadblock on the highway to prevent Panamanian reinforcements from reaching the tank farm, witnessed the evacuation of wounded by blacked-out ambulances.[20]*

At dawn, teams of Marines swept the area, looking for evidence to confirm the presence of the intruders. They did not find much. There were signs that bodies had been dragged through the area, and there was debris such as fresh, foreign-made battle dressings, scraps of camouflage utility uniforms, and chemical light sticks. But there were neither bodies nor spent ammunition shells. However, Marines later learned that the automatic weapons carried by some Panamanian soldiers were equipped with brass-catchers.[21]

During the next several days, Captain Valore and his Marines were debriefed extensively by Colonel Conley's staff, by Naval Investigative Service agents, and by Army intelligence specialists. Many of these official visitors also walked the ground over which the fight had occurred, repeating questions that had been asked by others. There was even an order from U.S. Army South for the Marines who fought in the jungle that night to submit to a urinalysis, which they did, with negative results.[22]

What the Marines told the debriefers was consistent, and detailed. Some of their testimony was also graphic. For example, when one of Cooper's Marines described his experiences during the firefight, he began by telling the debriefer how he had "gone to ground," and tried to make himself invisible. Nevertheless, one of the passing intruders spotted him at a distance of 8 to 10 feet away. As the intruder began to swing his weapon towards the Marine, the Marine shot him twice in the chest with his M-16. Upon impact, the intruder fell back and hit the ground so hard that "his legs flopped in the air."[23]

Even after the debriefings, doubts persisted. Some members of the joint task force staff acknowledged the threat, but wondered if the Marines had overreacted. More frustrating for the Marines were lingering questions about the facts. Matters took a turn for the worse when Southern Command's Public Affairs Officer failed to reject suggestions that the Marines had fired at shadows. Noriega exploited the opportunity, using his propaganda machine to plant stories about drug abuse among the Marines, implying that the events of 11 and 12 April were drug-induced hallucinations.[24]

Marine Forces Panama overcame the challenge to its credibility and professionalism. Colonel Conley and Captain Valore closed ranks behind their Marines, who reciprocated in kind. On the night of 13 April, when Captain Valore walked into one of the clubs at Rodman, all of the Marines present cheered him. He later said that this was one of the highlights of his career, and that he was moved by the bonding and mutual trust among men who were now combat veterans:

> The [pre-combat] tension was there; it was released. I [had] gained their credibility. They knew I was going to stand behind them, whatever it took... They could depend on me to look out for them, [and] not play politics or worry about other things... You have a mission to accomplish: you've got to take care of your Marines.[25]*

Valore's Marines stood down from the tank farm for a few days of rest and recreation at Rodman. A reinforced Army battalion took their place. On 14 April, Army sentries guarding the ammunition supply point challenged a small group of intruders, who responded with gunfire, and a patrol from the 3d Battalion, 7th Special Forces Group, operating west of Howard Air Force Base engaged another group of well-disciplined intruders. It was now clear that the Marines were not the only American soldiers to encounter intruders and to defend themselves.[26]

In the months that followed, there were occasional incidents, but nothing on the scale of the 12 April firefight. In retrospect, the aggressive Marine response appeared to have had the desired effect; the Marines had recognized the threat and responded accordingly. Information obtained from Panamanian officers following Operation Just

---

*Valore remarked to another Marine officer that, just before General Loeffke's arrival, he had ordered his reserve to flank the intruders. This movement did not take place due to Loeffke's intervention. (Maj William J. Philbin intvw with author, 23Jun94.)

*Marine Forces Panama officially reported that, despite the challenge to Marine credibility, morale was excellent. Headquarters Marine Corps took note of Valore's and Nevgloski's performance, awarding both the Bronze Star Medal with Combat "V." Sergeant Cooper received the Navy Commendation Medal. All of the Marines involved were awarded the combat action ribbon. (Maj Joseph P. Valore ltr to author, Nov93)

Cause, executed 20 months later in December 1989, confirmed what the Marines had said about the 12 April engagement. The intruders were almost certainly troops from the 7th Rifle Company, known as the "Macho de Monte," one of the few elite formations in the PDF, possibly reinforced by a few members of the Special Anti-terrorist Security Unit, the Panamanian equivalent of Delta Force. There is some evidence that one or more Cuban advisors may have accompanied the "Macho de Monte" on their 12 April sortie.[27]

Why did the intruders attack the fuel storage area? It is unlikely that the tanks themselves were the targets. Even when protected by a company of Marines, many of the tanks remained vulnerable, as did pipelines and other peripheral installations. If sabotage had been the mission, the intruders probably could have succeeded. Far more likely, considering the timing of the attack, a few days after the arrival of Marine reinforcements in-country, and supported by documents seized in December 1989, is the conclusion that the Marines themselves were the main target. The object was to embarrass and harass while testing the virility and competence of the Panamanian military elite. The anti-Marine propaganda campaign in the press had only just begun, and, as Noriega himself later commented: "Never before has there been a more effective laboratory for armed men. Nor has the opportunity [ever] been so propitious for training in low intensity conflict, as that which exists today in Panama."[28]

Recovery and Consolidation

The month that followed was a period of relative calm. It was one of diplomatic, rather than military activity, which gave Southern Command and the Marines an opportunity to prepare for future intrusions. The events of mid-April demonstrated the need to resolve ambiguities in the system of command and control. On 25 April, SouthCom activated Area of Operation Pacific West, and formalized the arrangement whereby Colonel Conley became the commander of all of the Marines in-country, as well as temporary attachments from the U.S. Army, including 293 soldiers of the 519th Military Police Battalion and Company B, 1st Battalion, 508th Infantry Regiment. Marine Forces Panama remained under the operational control of the joint task force, which meant that Colonel Conley would continue to report to General Loeffke, even though the normal chain of command would have had him to report to Admiral Gnecknow.[29]

In the meantime, Marines in the intelligence field analyzed the threat, and generated a collection plan in an attempt to predict future incursions, and the routes that they would take. Company I, after its two days of rest, went back on line for another 14 days. In addition to routine patrolling, day and night, the company improved and hardened six observation and listening posts in both the tank farm and ammunition supply point.* Throughout the period, the Marine forces conducted training exercises. First there were all-Marine, in-house exercises in the deployment of mobile reaction forces to trouble spots. The reaction forces included jeep-like HMMWVs carrying heavy machine guns and TOW missiles, a significant increment in the kind of firepower which had been used to good effect to suppress enemy fire on 12 April. Then there were a series of exercises with other services, codenamed Purple Storm for offensive evolutions and Purple Blitz for defensive evolutions and casualty evacuations.[30]

Not content with routine exercises and countermeasures, the Marines experimented with other means of accomplishing their mission. Air Force and Navy dog teams regularly joined the Marines on patrol. Army specialists helped the Marines set up a speaker system which, when triggered, informed intruders in Spanish that they were on U.S. Government property and that it was in their best interest to return to where they had come from. At night, the Air Force continued to fly AC-130 Specter missions in support of the Marines, using beacons and infrared devices to search for intruders. There also were attempts to intercept radio transmissions between groups of intruders. However, these attempts were unsuccessful, apparently because the intruders relied mainly on other means of communication.[31]

Neither War Nor Peace

The period of relative calm came to an end around 1900 on 19 July, when sensor activations alerted the Marines of Company L, 3d Battalion, 4th

---

*The ammunition supply point remained a Marine responsibility even though it was often patrolled by Army infantry attached to Marine Forces Panama.

Marines, which had relieved Company I in June. The sensors indicated a group of roughly 20 intruders, apparently gathering at a rallying point in the tank farm. Marine Forces Panama requested and received permission from the joint task force to insert blocking forces and capture the intruders. Accordingly, a Marine unit, which included the soldiers from Company B, 1st Battalion, 508th Infantry, began to move into place during the night–a tortuously slow proposition in the jungle under any conditions, and doubly so at night.[32]

Once again, Marine forces came away empty-handed. At first light, the appearance of an Army UH-1 Huey command helicopter apparently alerted the intruders to the risk of capture before all of the blocking forces were in place. The helicopter circled in the vicinity of the rallying point, enabling General Loeffke to monitor the operation. The intruders reacted by slipping out of the area by twos and threes. The only contact was a brief firefight between a Marine patrol and some of the intruders.[33]

The pattern of events in August and September was similar but less dramatic, with a number of sporadic sightings and brief firefights in both the tank farm and ammunition depot. Frustrations and tensions persisted. There were still no confirmed dead, and very little hard evidence to prove who the intruders were. It appeared to the Marines that many of their counterparts in the other services continued to doubt that the intruders were professionals.

In order to put the doubts to rest, and in general to bridge the cultural gap between the Marines and members of the other services at all levels, Colonel Conley developed an orientation program for Army and Air Force staff officers. He arranged for them to fire Marine weapons on the range and invited them to walk the lines in the tank farm at night, which usually included spending some time in a fighting hole watching and waiting. The reaction was positive, and helped to achieve Conley's goal. The visitors saw that the Marines were living in the field, under near combat conditions. Sandbags and gun pits reminded them of Vietnam.[34]*

*A similar initiative, aimed at maintaining the morale of Marines on the line, was "Staff and Officer Appreciation Day," when staff officers and non-commissioned officers would leave their desks and wield machetes to help clear vegetation in the tank farm. (Col Thomas W. Roberts intvw, 30Mar90, p. 38 [Oral History Collection, MCHC, Washington, D.C.])

But not everyone was convinced. For example, during the month of September, the Marines were able to capture on tape thermal images of the intruders by means of the UAS-11. The images clearly showed a number of individuals carrying AK-47s and wearing load-bearing equipment. But some officers at joint task force headquarters claimed that the images were not distinct enough to constitute proof.[35]

On 31 October, there was another major firefight, much like that of 12 April. Over a period of roughly two and one-half hours, there were seven separate incidents throughout the Arraijan Tank Farm. In five of the incidents, the intruders fired first, but did not hit any Marines. In the other two engagements, Marines from Company M, 3d Battalion, 8th Marines, which had replaced Company L, fired first with unknown results. After the last of the incidents, the outcome was still unclear. Reconstructing the chain of events in the light of day, the Marines developed two possible scenarios. The first assumed that a Panamanian Defense Force patrol had penetrated the depot, bumped into one observation post, and then, on its way out, run into another. One or more Panamanian patrols may have come to the rescue of the first patrol. The second was that the Panamanians had deployed a number of small patrols from the beginning. But, once again there were neither bodies nor spent shells.[36]

The repercussions of the October firefights were predictable. The Panamanian press exploited the Halloween theme, showing cartoons of Marines shooting at witches flying through the palm trees. For its part, Southern Command directed the joint task force to place limits on nighttime patrolling and the deployment of observation and listening posts in the jungle. The Marines had to count the number of rounds they fired, and then to account for them during lengthy debriefings. Limits also were placed on the number of Marines permitted in the tank farm at any one time. This often meant that there was only a reinforced platoon in Arraijan Tank Farm, leaving one platoon to patrol the ammunition depot and one in reserve, which gave the Marines an opportunity to rest and train. Later, the Marines were informed that, henceforth, their primary duty was to conduct training, and that their missions in the storage facilities were secondary. The theory behind the changes was that Marines attracted intruders. Therefore, if fewer Marines deployed, there would be fewer intruders and fewer incidents. It was almost an argument for leaving the tank farm and the ammunition supply point unguarded, which no one was prepared to do.[37]

The changes exacerbated the existing friction between Marine and Army commanders, each of whom had different concerns. The more senior Army commanders obeyed political dictates in order to minimize confrontation, while conditions on the ground dominated the thinking of Marine commanders, who never stopped pressing for more tactical freedom. Another result of the restrictions was, almost literally, a bunker mentality. Since the Marines could not go to the enemy, they had to wait for him to come to them. Once he appeared in front of their positions, the rules of engagement were such that the Marines either had to warn the intruder or leave it to him to initiate hostilities. One Marine officer remembered what it was like for the rifleman at his post:

> [The intruder] would come towards the position, and a Marine might lock on to [the intruder]... with his night vision system.... [B]ut the Marine... had to challenge the aggressor. So the agressor suddenly had the upper hand . . . because the aggressor knew that the Marines... were so well disciplined that... They would not open up unless the PDF...violated one of the rules of engagement.... [T]hen this bad guy shoots at you..., the blast from the enemy muzzle is going to momentarily... blind you and then... [you have] lost that opportunity because the aggressors are well trained and they hit the deck and back out of there.... [38]

Under the circumstances, many Marines felt that the best way to protect themselves was to improve their positions, which became ever more elaborate. By now, there was a system of bunkers and trench lines that covered parts of the tank farm and, to a lesser extent, the ammunition supply point. Finally it was clear that the pendulum had swung too far, and the new commander of Marine Forces Panama, Colonel Thomas W. Roberts, ordered virtually all of the positions in the fuel storage area razed. The joint task force concurred in the decision, as well as its corollary, that Marines return to light infantry tactics in both the tank farm and the ammunition depot.[39]

Roberts' reasoning was that the change would make life more risky for the intruder. Since most of the Marines would not be in fixed positions but patrolling or lying in ambush, the intruder would never know when or where he might encounter them. Roberts shared the JTF Panama's assumption that the Panamanian Defense Forces wanted to cause embarrassing incidents, but came to a different conclusion. If defense force commanders had to think twice before entering the storage facilities, there would not be many incidents. Colonel Roberts appeared to have been right; the net result of the change in tactics was a sharp decrease in the number of incidents.[40]

Operation Rabbit Hunt, which was launched in the ammunition depot in December, apparently had the same kind of deterrent effect. In what was essentially a cordon and search operation, Marine Forces Panama, accompanied by an entire Army airborne infantry battalion, the 1st Battalion, 508th Infantry, swept through the ammunition supply point during the early morning hours of 19 December, while Army military police and engineers patrolled the perimeter. No intruders were captured, due perhaps to possible security leaks from a joint command that was, not unnaturally, closely intertwined with Panamanian society. But, as were many of the other Marine initiatives, it was not a failure in that it helped to deter intruders, which remained as much a part of the mission as killing or catching them.[41]

Another significant event in December was the visit of the Chairman of the Joint Chiefs of Staff, Admiral William J. Crowe, USN, who toured the Arraijan Tank Farm and expressed shock at its size and dense vegetation. He found the lack of visibility troubling, noting that an intruder could approach unseen to within five yards of Marine positions. "I thought it was bad being inside a submarine," he said, "[but] [t]his is [really] horrible!"[42]

The Evolution of Training

As the months progressed, the training package instituted by Marine Forces Panama evolved to meet the demands of the situation. As most of the infantrymen who were destined to be deployed to Panama would be drawn from the 6th Marines, they were required to attend a tailored course of instruction before deploying. The course was approximately 38 days long. It stressed small unit leadership, patrolling, night fighting, and combat firing, day and night. Through trial and error, Marines learned that, when firing at night, more was not better. Results improved when they fired

When not on patrol, Marines honed their battle skills in live night-fire training. The training was aimed at creating a highly disciplined and technically proficient force.

one or two shots, evaluated the strike of the rounds, and made corrections. Finally, there were seemingly endless drills on the rules of engagement.[43]

The course was a success due in large part to the foresight and initiative of the regimental commander, Colonel Russell H. Sutton, who understood the need for constant, near real-time feedback from Panama to make the training fit the situation. To get that feedback, Colonel Sutton or one of his principal staff officers travelled to Panama on a regular basis.[44]

Training did not stop after the Marine arrived in Panama. First, everyone received a week of indoctrination and orientation at the Jungle Warfare Training Center at Fort Sherman in the Canal Zone. Then, even during the execution phase of the mission, there was more training. As often as every other night, front-line Marines cycled through three stations where they practiced using night vision goggles and combat firing techniques, before running through rules of engagement drills. During one training exercise, Marines moved tactically down an 800-meter jungle trail where they had to make split second "shoot or don't shoot" decisions while "minimizing" their own positions. Pop-up targets and flashbangs added realism. In other scenarios, Marines practiced firing at silhouettes at night. At the night vision station, trainers presented Marines with different scenarios, asking them to describe what they saw while wearing their goggles. The procedure at the rules of engagement station was similar. There were more than two dozen scenarios, requiring responses in Spanish and English. Over the course of time, most Marines internalized the rules, which changed somewhat over time but consistently directed Marines to use the minimum force necessary to counter threats. Under most circumstances during the deployment of Marine Forces Panama, if intruders were spotted, the Marines had to challenge them, and could take action only if the intruder committed a hostile act, which was defined as pointing a weapon at a Marine or attempting to flank his position.[45]

There was even a combat pistol course for Marines armed with the weapon, important because many of them had only fired their pistols at stationary targets at known distances in order to qualify. The course consisted of 10 rounds fired at eight targets, while kneeling and standing, with both strong and weak hands, as well as from behind a low wall, a high wall, through a window, and at point-blank range. To add stress, Marines had to complete the course in less than one minute. Every "pistol carrier" had to fire, or refire, the course before going to the field. To make sure that skills were not lost, the "pistol carrier" also had to fire a few rounds for familiarization almost every day.[46]

Most Marines became extremely proficient at shooting at night, far more so than their opponents who failed to wound or kill a Marine in the tank farm or ammunition supply point. On the other hand, Marines scored numerous hits on their opponents, not to mention the targets on the range, where nighttime scores of 15 out of 20 were the norm.[47]

Apart from the training in battle skills, there was continuing education for officers and staff noncommissioned officers. To develop a common vocabulary and to exercise tactical judgment, they conducted sand table exercises, many of them based on scenarios generated by Dr. William Robertson's Korean War classic, *Counterattack on the Nakong, 1950.*[48]

The result of the training was a highly disciplined and tactically and technically proficient force. Despite the restrictions, which were frustrating at times, there was painstaking adherence to the rules of engagement, which came easier because the Marines were confident in their own abilities. They knew that they could face most of the challenges that they were likely to encounter in Panama, which was not just another training area. Staying ready to execute a tangible mission helped to keep morale high, which in turn meant that the number of office hours to enforce discipline under the Uniform Code of Military Justice was very low.[49]

The New Year

Although there were incidents in January 1989, especially in the ammunition storage depot between the 6th and the 24th, the new year brought a period of relative calm, and then a virtual end to confrontation at both facilities. The situation was so quiet that a rumor that the Marines were going home in February made the rounds. However, the Joint Chiefs of Staff decided they were still needed in Panama, and the execute order never came.[50]

The reason why it was so quiet on the tank farm front was that the focus of activity had shifted to preparations for the election that Noriega had to permit in May. This meant that the PDF spent more of its energy on the streets and highways, which in turn created a new set of issues, that of freedom of movement. The problem was that the Panamanian government began to restrict the ownership of private vehicles by members of Southern Command, and to harass both individuals and convoys at roadblocks and checkpoints. One consequence of the shift in defense force tactics was that, among their counterparts in Southern Command, there was now more understanding and sympathy for what the Marines had endured.

Despite Noriega's best efforts, the ticket which he endorsed lost the 7 May election. Noriega's response was to send the paramilitary Dignity Battalions into the streets to attack his opponents, including the winning candidates for President and Vice President. The 10 May attack on Vice President-elect Guillermo Ford was immortalized by a press photographer who produced a sequence of photographs of Noriega's enforcers striking the defenseless politician with sticks, leaving him bleeding profusely onto his white shirt.

Overall, the period was similar to that of the previous March, and evoked a like response from American policy-makers. Within days of the election, additional Army and Marine reinforcements were on their way to Panama. Since freedom of movement was now one of the central issues, General Woerner wanted mobile forces that suited the terrain and the situation. He needed the kind of light armored vehicles (LAVs) available only in the Marine Corps inventory. The LAV was the state-of-the-art vehicle in its class, faster and more maneuverable than most of the alternatives. It was also formidable-looking, standing high on a eight-wheel chassis topped by a chiselled turret. The turret housed a 25mm "Bushmaster" chain gun and 7.62mm machine gun. Depending on its configuration, most vehicles could also carry eight infantrymen.

Given General Woerner's preference, it was no surprise that additional Marine reinforcements were drawn from Company A, 2d Light Amphibious Infantry (LAI) Battalion, equipped with LAVs, or tanquitos (little tanks), as the Panamanians soon labelled them. Accompanied by a battalion command group and lifted into Howard Air Force Base by six Air Force C-141 Starlifter and three C-5 Galaxy aircraft, the tanquitos of Company A arrived in country on 12 May. The contingent comprised 14 light assault variants (LAV-25s), two logistics variants (LAV-Ls), and one command and control variant, as well as 10 four-man scout teams to protect the LAV-25s.[51]

The company's primary mission was to participate in operations to ensure that U.S. forces in Panama enjoyed freedom of movement in and around the Canal Zone, which was guaranteed under the Carter-Torrijos Treaty in order for them to defend the Canal. The security operations were known as "Sand Fleas." Participation in this series of operations signalled a change for Marine Force Panama, which had been largely confined to the boundaries of Rodman and neighboring military installations. It can be argued that it was the LAVs that transformed Marine Forces Panama from a reinforced security force into a maneuver force.

Within days of their arrival, the LAVs seemed to be everywhere at once, an intentional tactic designed to keep the Panamanian Defense Forces off balance. Their effect on the PDF was clear in incidents like that of 23 May, when Panamanian military police stopped a column of LAVs returning to Rodman from a routine patrol, claiming that the LAVs were machines of war and therefore required an escort. The LAI platoon commander replied that, by stopping his vehicles, the Panamanians had violated his right to freedom of movement. He gave the policemen a card which spelled out American treaty rights and allowed them two minutes to consult with their superiors before he would "carry out his mission." After making a show of attempting to call a superior, the Panamanian soldiers apologized before permitting the LAVs to continue on their way.[52]

A few days later, in Operation Big Show, the LAVs swam the Canal from west to east in an exercise designed to secure U.S. housing and other facilities at Fort Amador. The tactic of crossing the Canal at will caught the Panamanians off guard, and was successful enough to be repeated on 20, 23, and 26 June in similar operations. One month later, on 22 July, the company executed Operation Hippocrates, designed to test their ability to create a diversion on the west bank of the Canal that would make it difficult for the PDF to send reinforcements to the capital on the east bank. That same day the Joint Chiefs issued National Security Directive 17, which formalized what was already fact–giving commanders in Panama the authority to conduct operations to destabilize the PDF, ensure freedom of movement for U.S. forces, and rehearse and reconnoiter for follow-on operations.[53]

One of the more successful Sand Flea operations was Westward Ho, yet another demonstration of the LAVs usefulness in low-intensity conflicts. The operation began early on 7 August. Two platoons and a command element conducted a route reconnaissance along the southwestern border of the Canal Zone. Panamanian Defense Force elements detained one of the LAVs near the town of Arraijan. A full platoon of LAVs went to the rescue, while Army helicopters monitored defense force activity from the air. Another platoon of LAVs stood by to block the routes along which the Panamanians could send reinforcements. Meanwhile, the command element lured a squad-sized element of the PDF onto American-controlled territory, where Marines detained its members and confiscated approximately one dozen weapons and thousands of rounds of ammunition.

Marine light armored vehicles cross the Bridge of the Americas during one of several freedom of movement exercises. The exercises ensured that American forces enjoyed free movement in and around the Canal Zone and kept the Panamanian Defense Forces off balance.

Major Manuel Sieiro, Noriega's brother-in-law, was among the detainees, who were released after being disarmed and photographed.[54]

Westward Ho was a typical operation in that LAI operations were the focus of effort for Marine Forces Panama during the summer and fall of 1989. It was also typical in that the PDF failed to impede the movement of the Marine vehicles. What was less typical was the Marines' propaganda coup. While most Sand Flea operations clearly demonstrated the futility of harassing mobile and disciplined troops, the detention of Noriega's brother-in-law and his cohorts was an unexpected windfall which increased Marine self-confidence. This was important in an uncertain environment where it remained difficult to measure success.[55]

Over time, LAI operations became more ambitious and ranged farther afield, as in the case of Operation Lola, a 300-kilometer motorized reconnaissance which began on 23 September, a little more than one month after Company B, 2d LAI Battalion replaced Company A, which rotated home. The operation took the company west through jungle and rolling terrain to the town of Arenosa. The local populace welcomed the Marines in what was generally a show of friendly curiosity despite the open hostility of PDF "escorts," who followed the LAVs in chase cars.[56]*

Despite the focus on mobile operations, the LAI routine was not very different from that of the rifle companies that continued to serve in Panama. The company generally was on a schedule of three-day rotations, with one platoon on patrol, another in training, and the third in maintenance. Training emphasized 25mm gunnery, small arms marksmanship, and rules of engagement. There were live-fire exercises virtually every day. What was somewhat different from the training for infantry Marines was the roadblock training. While the scout platoon practiced crowd control formations, other LAI Marines developed techniques for clearing both hasty and deliberate roadblocks.[57]

The October Coup and its Aftermath

The LAI company was one of the few Marine units which played a role in the events of 3 October, when Panamanian military dissidents under Major Moises Giroldi attempted to overthrow Noriega, but succeeded only in holding him prisoner for a few hours in the PDF Commandancia, its headquarters in the capital. The day before, Marine Forces Panama was placed on standby, having received orders to prepare to block the approaches to Panama City from the west, where the 6th and 7th PDF Companies had their bases. The order to execute finally came around 1100 on the 3d, while the outcome was still in doubt. While other Marine units moved to staging areas, 33 FAST Marines under First Lieutenant Robert F. Killackey responded. A detachment of Army engineers and a reaction force of LAI Marines in light armored vehicles provided support. Killackey's Marines established a roadblock approximately two kilometers west of the tank farm, where the Pan-American Highway intersected a north-south highway. The engineers placed obstacles, while the LAI Marines waited in reserve nearby. There was air support on station. Just to be sure, Killackey had even taken the trouble to plot artillery targets.[58]

Tension was high. The Marines knew that they could face up to two companies of PDF soldiers, possibly supported by Cadillac Gage V-150 armored cars and V-300 armored personnel carriers, some armed with 90mm guns. But it soon became clear that the greater threat was from the east. Hundreds of civilians were fleeing the capital on account of the coup. At times near panic, the crowd pressed against the Marine positions. The Marines held the crowd back as much by their professional and aggressive appearance as by their obstacles. They did not need to fire any shots, or come into physical contact with any of the would-be refugees.[59]

Although the perimeter held, two things became clear with time. One was that, if the Marines had to open fire for any reason, civilian casualties would be high. The second was that PDF reinforcements were not travelling by road. Ultimately, Noriega was rescued by troops airlifted into the capital. That being the case, Killackey's Marines were withdrawn after being in place for approximately two hours. Once again, restraint and common sense had served the Marines well.

The coup attempt prompted Southern Command and the Joint Chiefs of Staff to reevalu-

---

*The PDF was able to deploy "escorts" since advance warning of impending movements generally was required by the Canal Treaty. In this instance, the PDF "escorts" did not hinder American freedom of movement.

ate existing contingency plans for Panama. Collectively known as Prayer Book, they included plans for neutralizing the PDF, defending the Canal, and restoring law and order. But they were primarily defensive in nature. They lacked the range of options, specifically offensive options, that the new Chairman of the Joint Chiefs of Staff, General Colin L. Powell, and the new SouthCom commander, General Maxwell R. Thurman, wanted. In particular, they wanted to change Blue Spoon, the plan for neutralizing the PDF, in order to provide a capability to respond virtually without notice to unforeseen contingencies and to combine conventional and special forces, such as Army Rangers and paratroopers, for night operations. They also wanted to be able to do more than simply neutralize the PDF. After 3 October it was clear that simply eliminating Noriega and replacing him with another Panamanian military officer would have little effect on the situation. The PDF was, after all, his creature.

The need for change meant a larger role for the XVIII Airborne Corps under Lieutenant General Carl W. Stiner, USA. His command was the U.S. Army's crisis action team which any of the regional commanders-in-chief could task. Able to call on some 85,000 troops at posts throughout the United States, and to deploy forces on short notice, it had a large and competent operations staff, as well as some of the best troops and equipment in the U.S. arsenal. Stiner himself was an experienced and aggressive field commander who already had a contingency plan for Panama on the shelf. But now Powell and Thurman decided to enlarge Stiner's role. His staff would take the lead in planning operations in Panama and, in the event of hostilities, he would assume overall command, becoming Commander-in-Chief, Joint Task Force South.

The final version of Stiner's operations plan, known as XVIII Airborne Corps Op Plan 90-2, emphasized overwhelming force, maximum surprise, and centralized control. In line with guidance from Washington, it also emphasized the need for restraint in the application of force in order to limit collateral damage and civilian casualties. The principal missions under the plan were to protect the canal, defend treaty rights, remove Noriega from power, and create a substitute for the PDF to protect a democratically-elected government in Panama.

Stiner's plan of attack called for operations to begin with a number of simultaneous attacks throughout the country. The primary mission for Task Force Semper Fi, the designation given Marine Forces Panama and its attached units under 90-2, was to support the attacks by defending the western approaches to Panama City, including the Bridge of the Americas which connected the east and west sides of the Pan-American Highway over the Canal. Since Marine forces were already perfectly positioned, the planners needed only to make it official. Although not explicit in all versions of the plan, other missions assigned included: protecting U.S. facilities at Howard, Rodman, and neighboring Fort Kobbe; seizing the port of Vaca Monte; and neutralizing all PDF threats in its assigned zone, which included bases and stations in the towns of Arraijan and Veracruz. The security of Howard, through which troops and equipment flowed into Panama, eventually became Marine command's focus of effort.[60]

In October and November, General Stiner and his staff held a series of meetings in Panama with subordinate commanders to discuss the plan. The commander of Marine Forces Panama, Colonel Charles E. Richardson; his operations officer, Lieutenant Colonel Michael J. Franks; and liaison officer, Major Bron N. Madrigan, represented the Marine position with energy and imagination. They made it clear that 90-2 represented a substantial change and expansion of the original Marine mission in Panama. Under 90-2, the Marine command could become responsible for an area which embraced literally hundreds of square kilometers. Remembering that it had originally deployed as the advance force for a Marine expeditionary brigade, and that Blue Spoon had contained provisions for a MEB, Franks suggested in early October that the MEB be written into 90-2. When it became clear that this would not happen, both Richardson and Franks repeatedly requested that at least one Marine battalion deploy to Panama to augment the task force. Ultimately, General Stiner decided against deploying additional Marine forces. However, he did agree to provide a battalion of the 82d Airborne Division no later than 36 hours after H-Hour, in addition to the 536th Engineer Battalion and the 534th Military Police Company. Later in the planning cycle, Stiner's staff decided to substitute a battalion of the 7th Infantry Division (Light) for the battalion of the 82d Airborne Division.[61]

In the meantime, the LAI Marines continued to conduct Sand Flea operations. Given the heightened possibility of a large-scale operation against Noriega and the PDF, these operations were more important than ever. While Panamanian military responses may have slowed somewhat due to the purge which followed the coup attempt, there were still difficult times. On 31 October, for example, Company B encountered a roadblock at La Chorerra, several kilometers west of the Canal Zone along the Pan-American Highway, during the execution of Operation Chisum. The Marines were disappointed when JTF Panama ordered them to turn around rather than accept the risk of collateral damage which attempts to breach inevitably entailed.[62]

On 22 November 1989, during Operation Rough Rider, matters were refreshingly different for the next company of LAI Marines to serve in Panama--Company D, 2d LAI Battalion, under Captain Gerald H. Gaskins. Planned to familiarize the fresh company with the towns of Nuevo Emperador, Nuevo Guarare, Vista Allegre, and Arraijan, the operation called for the deployment of 12 LAVs and three HMMWVs, supported by an Army psychological operations detachment and two treaty affairs officers, as well as an OH-58 Pave Low helicopter. On the outskirts of Vista Allegre, the leading LAV reported a crowd waving Panamanian flags and several banners in front of a roadblock. At that point, the lead element halted 75 meters from the roadblock, while the rear element set up a counterblock to prevent any surprise from the rear. When a shouting match erupted between the treaty affairs officers and the crowd, Captain Gaskins requested permission to warn the crowd that, in five minutes, the LAVs would breach the roadblock. Almost immediately, Marine Forces Panama granted permission. Captain Gaskins ordered the treaty affairs officers to reboard their vehicles and the company to move. He described what happened next:

> The agitators realized what our intentions were, and several vehicles were hit by rocks. The company . . . collapsed into a single file except for the rear element which blocked traffic until every vehicle was through. One Panamanian rammed an LAV-L with a pickup truck . . . . [A] female agitator tried to block our movement with her body. She fell backwards, feet in the air, flipping over another . . . vehicle. The agitators were shocked, and began beating our vehicles with their fists, flag poles, and anything else they could [lay their hands on] . . . . On the other side of the roadblock was another crowd, this one friendly, cheering . . . what we had done. . . .[63]

Five minutes later the company was again moving down the open road. Soon, however, it encountered two more roadblocks. It breached the first without stopping. The second, near the town of Rio Potrero, was built around a farmer's pickup truck. The PDF had apparently forced him and his neighbors to set it up. When the treaty affairs officers told the farmer that the Marines had simply pushed vehicles out of their way at two previous roadblocks, he and his friends redirected their anger at the Panamanian troops and drove the little truck away.[64]

Block and counterblock technology continued to evolve after Rough Rider. The Panamanians staged junkyard vehicles at key locations to create roadblocks, and stacked old tires in front of them, ready to set them on fire. The Marines acquired grappling hooks and fire extinguishers to remove the threat, and prepared for more serious threats, such as Molotov cocktails, by issuing M-14 rifles to two trained snipers in the company.[65]

Setting the Stage for Just Cause

Although Sand Flea operations continued throughout December, the Marines spent more and more time planning and rehearsing for contingencies under Plan 90-2. Marine counterintelligence operatives and interrogator-translators continued to collect data on potential threats and targets of all kinds, including specific buildings and individual defense force personalities. Their reports, often the fruit of long hours on the streets and in the barrios around Rodman, placed a wealth of information at the disposal of Colonel Richardson's staff. With that knowledge in hand, they were able to conduct realistic planning exercises, and prepare for operations against a range of potential targets. As part of the process, operations officers drafted sets of orders, written for both offensive and defensive scenarios, which were distributed to all companies. First Lieutenant Kenneth M. DeTreux recalled that at least once a week company commanders briefed their officers on recent developments and pointed out possible changes in the orders that could result from them. Then the officers would conduct map exercises. Colonel Richardson also conducted frequent sand table

Photo by CWO-3 Charles W. Rowe, Jr.

Light Amphibious Infantry Battalion Marines train for urban warfare. In the event of conflict the Marines were to be prepared to secure a number of Panamanian Defense Force facilities.

exercises for the same purpose.[66]

Company D, 2d LAI Battalion, prepared to secure a nearby PDF facility, Directorate of Traffic and Transportation (DNTT) Station No. 2, in Arraijan, and, to a limited extent, trained for military operations in urban areas. Although none of the Marines, either in the LAI or the infantry company, received a great deal of urban training, at least the LAI scout platoon practiced, as a unit, for operations that went beyond the simple mission of each fire team protecting its own LAV. Other LAI Marines worked with the Army's Task Force Gator, whose objective was to neutralize the Commandancia. After witnessing the accuracy of LAV gunnery, and conducting extensive tests, including firing into windows at ranges up to 2,000 meters, Army planners concluded that the LAV was the most effective weapon in the JTF Panama inventory for engaging targets inside the PDF Headquarters, a complex of 15 buildings including a three-story fortress-like office building. The result was that one LAI platoon was ordered to join Task Force Gator. To prepare for another possible mission in downtown Panama City, FAST Marines practiced operations to reinforce the detachment of Marine Security Guards at the American Embassy, which was near the Commandancia.[67]

It bears emphasis that virtually all Marines were involved in preparations for implementing 90-2. Colonel Richardson realized that, given the ambitious nature of the mission, it was likely that he would commit Marines from administrative and combat service support units to combat, that is, commit Marines whose primary job was to support the Marines on the front lines. They practiced basic combat skills and literally fired their weapons day and night until he was satisfied with their proficiency.[68]

The added responsibility was especially welcome to one Marine, Second Lieutenant Thomas H. Goessman, who had joined the Marine Corps to become an infantry officer but, by the luck of the draw at the Basic School, became a supply officer. He was elated when it dawned on him that he was likely to see more action than many of his brother officers who were serving in combat arms billets outside Panama. Most, if not all, Marines in Panama shared Goessman's attitude; they wanted to be there and to share in conducting the operation.[69]

The tempo of events increased significantly after 16 December. That day, Marine First Lieutenant Robert Paz, who was assigned to the operations section at SouthCom and therefore not subject to the precautionary Marine command regulation against liberty in Panama City, drove with three other officers to have dinner in the capital. They lost their way, and found themselves at a PDF checkpoint near the Commandancia. Panamanian soldiers carrying AK-47 rifles shouted threats and attempted to drag them out of their car. At that point, the young officers decided to drive away. At least one of the PDF soldiers opened fire with an AK-47, and hit the car. Paz, who was sitting in the back seat, was struck by a round which entered through the trunk. He was pronounced dead a few minutes later at Gorgas Army Community Hospital. Navy Lieutenant Adam J. Curtis and his wife, Bonnie, witnessed the incident, having been stopped at the same checkpoint. Shortly after the shooting, they themselves were taken to the

Commandancia, where PDF soldiers beat the lieutenant and threatened his wife. They were released a few hours later.[70*]

American military leaders in Washington and Panama reacted strongly to the incidents, especially since Noriega had declared on 15 December that a state of war existed between the United States and Panama. It now appeared that American servicemen were at risk in Panama, and that the PDF was out of control. The decision was relatively easy. To do nothing but protest was unacceptable. To remove Noriega and leave the PDF intact was not any more attractive now than it had been in October, especially since the PDF appeared to have acted more or less on its own on 16 December. The only option remaining was to implement 90-2 contingency plans against both Noriega and the PDF, which the Joint Chiefs of Staff discussed when they gathered at General Powell's quarters in Washington on Sunday, 17 December.[71]

During the meeting, the Commandant of the Marine Corps, General Alfred M. Gray, offered to commit the 11th Marine Expeditionary Unit (MEU), Special Operations Capable (SOC), to Panama. He pointed out that the unit, a composite Marine Air- Ground Task Force built around a battalion landing team of infantry, was thoroughly trained and ready. It was already embarked on board ship and, unlike airborne units, carried 30 days worth of supplies. Currently between Hawaii and California, it could be off Panama within a few days.** However, the consensus among the Joint Chiefs was that the United States needed to react as soon as possible, and that the approximately 24,000 troops already earmarked for the operation under Contingency Plan 90-2 would be more than enough to ensure quick success. With the Joint Chiefs in full agreement, General Powell secured the President's approval to proceed. H-Hour was set for 0100 on 20 December.

The decision to execute 90-2, now renamed Just Cause by the Joint Chiefs of Staff, was closely held, and even after being briefed by the current JTF Panama commander, Major General Marc A. Cisneros, USA, on the night of 18 December, Colonel Richardson could not inform his staff. Returning from the meeting, all he said to them was that Lieutenant Paz had not died in vain, and then set about making sure that his Marines were as prepared as they could be without knowing the plans. This meant that 19 December became a maintenance and logistics day. Captain Gaskins told his officers that, until further notice, their appointed place of duty was the maintenance ramp. While the LAI officers were at the maintenance ramp, Colonel Richardson attended another meeting at JTF Panama headquarters, and received permission to brief his officers at 1800, and to allow them to brief their troops three hours later.[72]

Even though the Marine mission did not coincide exactly with expectations, and even though the decision to execute at night came as a surprise, the objectives and tasks were by and large familiar, ones which the Marines had anticipated in their training. At 1800 on the 19th, Colonel Richardson briefed individual units on their assigned missions. Company D, 2d LAI Battalion (Minus) (Reinforced), was to attack and seize DNTT Station No. 2, attack and seize the PDF station in Arraijan, and screen to the west of Arraijan. Company D's 2d Platoon was to support Task Force Gator. Company I, 3d Battalion, 6th Marines (Reinforced), was to establish a roadblock south of Howard AFB, attack and seize the PDF station in Veracruz, and maintain security in the tank farm. When relieved by the Marine Security Force Company, it was to occupy the high ground west of Howard, and screen the approaches to the airfield. The reinforced detachment from BSSG-6 was to form a provisional rifle platoon, seize and hold the Bridge of the Americas, and provide combat service support to Marine command. The Marine Security Force Company was to relieve Company I in the tank farm, follow in trace of Company D and establish a roadblock west of the fuel storage facility on the Pan-American Highway, and maintain security at other U.S. Navy installations. The FAST Platoon, in addition to serving as the command's reserve, would be prepared, on order, to reinforce the Marine

*At least one writer has raised questions about the incident, suggesting that Lieutenant Paz and the other officers in the car may have been on some kind of mission, possibly of their own making. The driver of the vehicle, Captain Richard E. Haddad, was also a Marine Corps officer assigned to Southern Command. See John G. Roos, "Did President Bush jump the gun in ordering the invasion of Panama?" Armed Forces Journal International, Sep92, pp. 10, 14.

** Three amphibious group ships with the embarked Marine Expeditionary Unit arrived off Camp Pendleton on 19 December, while the fourth arrived in San Diego and offloaded on 22 December.

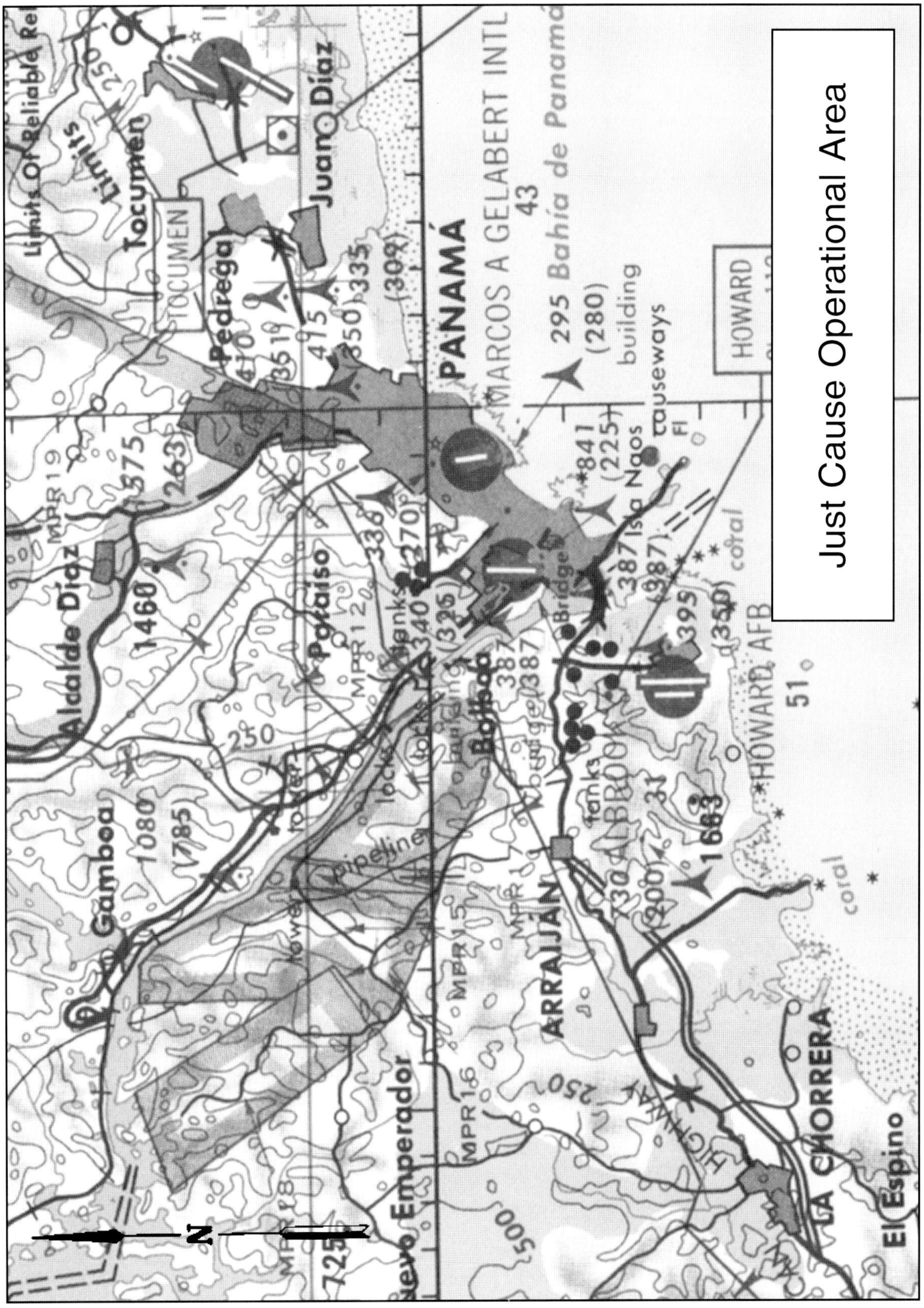

Just Cause Operational Area

Photo by Sgt Robert C. Jenks

Marine Forces commander Col Charles E. Richardson, center left, paces the floor of the combat operations center two hours before Operation Just Cause is launched.

Detachment at the U.S. Embassy and provide a close-quarter battle element for Company D.

When Richardson briefed the mission, he stressed security of operations and the use of minimum force. He told his commanders to avoid damaging historical sites and public works, and to give the PDF the opportunity to surrender. No one was to forget that the Panamanian people were not the target. But he did not want anyone to neglect the mission, concluding with the words "attack all targets," including the PDF, the traffic police, and the Dignity Battalions.[73]

When Major Robert B. Neller, commander of the security force since August 1988, briefed his Marines, he spoke for many other Marine commanders when he said that this was the moment of truth, one for which many of them had been waiting. He told his Marines that it was all right to be scared; that was normal. All they had to do was to do their jobs and take care of each other. The Marines were quiet and serious; there was no cheering. But there was commitment. For example, three Marines who had already checked out of the command and were waiting to fly home quietly drew their weapons and went back to work.[74]

Combat

Even before H-Hour, Marines went into action. At approximately 0040, Captain Gaskins received a report that Panamanian V-300 armored personnel carriers were on the move in the vicinity of Rodman. He assumed that they were heading for Panama City, and, mindful of his mission, ordered his Marines to begin moving to their blocking positions. At approximately 0050, 13 Company D LAVs moved out, loaded with the scout platoon, under Sergeant Thomas P. Bernius, and 17 FAST Marines. As the LAVs approached DNTT Station No. 2, the PDF opened fire with small arms from nearby bunkers. While 1st Platoon proceeded towards its objective at Arraijan despite the fire, 3d platoon turned its attention on the station. With rounds pinging off their armor, three of its LAVs burst into the station compound, firing their M-60 coaxial machine guns; they did not fire their main guns in order to limit collateral damage. Two of the vehicles stopped about 20 meters from the station and continued to fire while Sergeant Bernius and a squad of scouts under Corporal Garreth C. Isaak dismounted and ran up to the building. Bernius and Isaak forced the front door open with buckshot rounds from M-203 grenade launchers.[75]

Once inside the dimly lit building, the Marines faced a confusing tableau of muzzle flashes and ricochets. They could hear the rounds strike the walls, and they defended themselves by firing in the direction of the flashes. Working their way down the central passageway, they proceeded to clear the rooms that opened from it. When they reached the end of the passageway, a PDF soldier ran into one of the rooms and closed the door. During a lull in the firing, Corporal Isaak kicked the door open and, firing his weapon, started into the room. He was hit while still in the doorway, and fell back into the passageway. In the next instant, one of the other Marines fired buckshot into the room, and, seconds later, yet another Marine fired his squad's automatic weapon in the same direction. Then Sergeant Bernius ordered one of his men to throw a grenade into the room. The scouts were under orders to use grenades sparingly as another part of the effort to limit collateral damage. Finally, he ordered the squad to carry Corporal Isaak outside. They left Isaak, who probably had already died, with the corpsman on the logistics LAV, and continued on towards Arraijan in accordance with their orders from Captain Gaskins to spend no more than 10 minutes at the station. A few minutes later, a team of four Marines from Company D Headquarters again swept through the DNTT Station. In the room where Corporal Isaak was hit, they found a severely wounded PDF soldier, who reached for his weapon when he saw the Marines. One of the covering Marines shot and killed him. They cleared the rest of the building without incident. The final toll for the operation against the station was one Marine killed and one wounded. One Panamanian was killed, another wounded, and three were detained. For the Marines, it was the single most costly engagement during Just Cause; Corporal Isaak was the only Marine killed in Panama during the fighting.

Corporal Isaak's death brought home the seriousness of the enterprise to the LAI Marines, virtually all of whom learned of it within minutes. Henceforth, they were more conscious of their security. The LAVs buttoned up, with only the driver and the vehicle commander exposed, and the vehicles ran with blackout lights. There was also less of an inclination to err on the side of restraint. When consulted, the Marine command's operations officer, Lieutenant Colonel Franks, consistently made it clear that the commander's intent was to limit collateral damage, but only to the extent possible, and certainly not at the cost of any Marine's life.[76]

While moving west on the Pan-American Highway, 1st Platoon (Reinforced), the platoon that had not stopped to neutralize the DNTT Station, encountered a roadblock outside Arraijan at approximately 0120. The roadblock was built around two gasoline tank trucks. Captain Gaskins stopped his LAVs about 1,500 meters from the roadblock in order to guard against a possible rocket-propelled grenade attack. The scouts deployed to reconnoiter, and reported that there were 20 to 30 armed personnel around the tankers. In the meantime, 3d Platoon arrived and established an overwatch position, while FAST Marines provided forward, rear, and flank security. Gaskins hesitated to fire on the tankers because of the potentially disastrous consequences but, at 0210, decided that he had little choice. He requested permission to do so from Marine Force Panama. Lieutenant Colonel Franks, monitoring radios in the operations section, gave him that permission. The tankers turned out to be empty, and the rounds from the armored vehicles' main guns dispersed the opposition without incident.[77]

Within minutes, lst Platoon arrived at its next target, the PDF substation in Arraijan. Fifteen scouts dismounted and took cover against the wall surrounding the small two-story building. Finding both the wall and the building impervious to small arms fire, they pulled back, and permitted the vehicles' main guns to create openings for them. The scouts ran through the openings, and, learning from Corporal Isaak's experience, used fragmentation grenades to secure the building, which ultimately burst into flames. Within the hour, it was little more than a smoking ruin.[78]

Meanwhile, Marines of 3d Platoon proceeded to another target in Arraijan, the local headquarters of Noriega's political party. They, too, found it difficult to force entry. All the doors and windows were protected by grillwork that was locked into place. The situation called for a different method of operations. During the attack on the PDF substation, Captain Gaskins had persuaded Lieutenant Colonel Franks to allow him to use the FAST Marines, who had been following in trace and were specialists in forcible entry techniques. They now advanced on the building. While the LAI Marines and the LAVs provided perimeter security, the FAST Marines used shotguns to break the locks, and then warned the occupants of the building in Spanish. Seconds later, they threw a stun grenade through the door, and then entered. A Panamanian, visibly shaken by the effects of the grenade but otherwise uninjured, staggered out into the night air. The FAST Marines proceeded to secure the building without causing any casualties, thereby proving the value of their extensive training in close-quarter battle techniques. As a result, they now became the recognized "indoor" specialists, which meant that Gaskins and other commanders began to use them as the weapon of choice for clearing buildings.[79]

Shortly after 0400, its targets having been secured, the company almost literally circled the wagons to rest before proceeding around 0730 to search the homes of Rigoberto Paredes and Lieutenant Reuben Dario Serrano, both prominent Noriega supporters. Not surprisingly, neither was home.[80]

By dawn on 20 December, it was clear that the Marine units were accomplishing their principal mission, that of blocking the western approaches to the capital. As early as H-Hour itself, 0100, Marine Security Force elements had set up the roadblock between the tank farm and Arraijan with the help of Army engineers, who provided earth-moving equipment to place in the roadway. Security Force Marines had also assumed the mission of providing security in the fuel storage facility from Company K, which had begun its deployment to positions further west. A contingency platoon from the Marine Force's command element had secured the northern roadblock, located north of Rodman near Miraflores Lock on the Canal. The support group detachment, which was organized into a small task force of 25 Marines reinforced by Army engineers, had secured the Bridge of the Americas by 0106. The detachment commander, Captain William S. "Scott" Aitken, described the seizure in graphic terms:

> H-Hour came . . . .The skies were filled with Army helicopters heading toward Panama City and the sound of AC-130 gunships pounding the PDF headquarters . . . . As the [BSSG] convoy approached the bridge, it was exposed to firing . . . [From] across the canal. Fortunately the firing was not accurate . . . . [By now] the sky was orange and full of flames. The Marines quickly got barbed wire in place and the U.S. Army engineers . . . moved their large earth movers into position to completely close off the highway . . . . [W]ithin minutes the Bridge of the Americas was secure . . . .

The Marines and engineers on the bridge had a ringside seat for the fighting to their front, but, apart from hundreds, and later thousands, of Panamanians who wanted to leave the city on account of the fighting, no one challenged the detachment's right to the bridge.[81]

Some of the firing which the support group Marines heard and saw was initiated by their brother Marines in the four LAVs detached to Task Force Gator. The LAI Marines were providing covering fire for the Army assault on the Commandancia. They fired more than one hundred 25mm rounds with pinpoint accuracy through specific windows in the large building, while AC-130 Specter gunships fired at it from the air with 105mm cannon and Gatling guns.[82]

While most of the Marines in Panama spent the night on the move, one detachment of Marines in the capital had as its mission to stay put. The eight-man Marine Security Guard Detachment under Staff Sergeant Michael S. Pellow at the American Embassy withstood a kind of siege during the early morning hours of 20 December with little assistance or support. The FAST detachment which had prepared to reinforce the Marines inside the embassy was busy elsewhere, and the only semblance of support from fellow Marines came from two of the four LAVs attached to Task Force Gator, positioned on the streets near the embassy after supporting the attack on the Commandancia But the presence of the LAV detachment did not make much of a difference to Pellow's Marines, whose posts were inside the embassy compound. For all intents and purposes, they were on their own throughout the night.[83]

From 0030 on, they could hear firing, and, at times, see PDF soldiers run by the posts in the embassy. Pellow ordered his Marines not to fire at the passing targets, certainly a prudent decision, since the PDF showed little interest in the embassy before 0122, when three rocket-propelled grenades (RPGs), fired from a position in front of the embassy, struck the compound. When the impact occurred, Pellow was reporting by telephone to his company commander in Miami, and had to cut the conversation short. Believing that his Marines were about to make a last stand, the company commander told Pellow to "take as many of them as you can with you." Pellow ordered his Marines to load solid shot in their shotguns, and to fire at anyone carrying an RPG. (It was not Pellow's decision to arm his Marines with shotguns. State Department policy determined the table of equipment for Marine Security Guard detachments.) They did not acquire any targets before the next salvo of three RPGs, this time fired from a position behind the embassy, struck about one hour later. Their only potential target was a PDF officer with a radio that one of the Marines spotted near the front gate. To make sure that the officer did not order another attack, possibly by infantry, Pellow ordered his Marines to fire tear gas at the officer. He was "dusted" with a generous dose. Later, they fired more rounds of tear gas at a park across the street which the PDF could have used to stage troops for an assault. But there was no assault and, when morning came, the Marines were exhausted but uninjured and, in at least one case, covered with dust from a wall hit by an RPG round. Looking back on their performance after the fact, General Thurman commented that an infantry company of lesser men could not have done as well as the eight embassy Marines.[84]

While the order to execute Just Cause changed the mission for most Marines, there was relatively little change in the day-to-day routine for the Marines of Company K. It was, again, the luck of the draw. The company's missions, those of securing Veracruz and protecting the western flank of Howard Air Force Base, took them to a ramshackle seaside community and a string of hills covered with jungle which stretched in a five kilometer arc from Sierra Minon to Cerro Galera. For the average Marine, duty in those hills was not much different from duty in the tank farm or ammunition supply point.

The 2d Platoon, under Second Lieutenant Jeffrey R. Chessani, received the most challenging assignment, that of securing Veracruz. As his platoon approached in the town, the night seemed so unnaturally quiet that Chessani wondered if he was leading his Marines into an ambush. At approximately 0140, when they arrived at their first objective, the Panamanian Defense Force station in Veracruz, there was no one to be found. The station was strewn with clothing, equipment, and documents of all kinds, including passports, shot records, school certificates, and check books. There was even a half-eaten meal, further proof that the occupants had left in a hurry. After securing the station, the platoon withdrew from the

Sgt George Gamma, a Marine sniper, lays down protective fire on the entrance to the Panamanian Defense Force regional headquarters at La Chorrera in preparation for the Marine assault on the compound.

Photo by Sgt Robert C. Jenks

town to establish a blocking position and to patrol the area.[85]

In the meantime, 1st and 3d Platoons moved into their blocking positions in the nearby hills, where they established patrol bases. They began a routine that would vary little for the duration of the operation. Between daybreak on 20 December 1989 and 5 January 1990, they spent most of their time quietly patrolling, setting practice ambushes, searching farms, and detaining suspects. Although they received fire from time to time, they themselves did not return fire, either because they did not have a definite target or because they could not fire without running the risk of hitting nearby civilians.[86]

La Chorrera

Early on the morning of 20 December, Marine Forces Panama received a verbal order from JTF South to assault and secure the headquarters of the 10th Military Zone in La Chorrera–Panama's third largest city with a population of 80,000. It was an unexpected mission that would take Marine units west of their originally assigned area of operations. Nevertheless, it was also a logical objective for the Marines since it was the headquarters for most of the PDF units in that area. Colonel Richardson decided to assign the mission to Company D and its attachments from the FAST Platoon.[87]

At 1530, the company proceeded west on the Pan-American Highway. Ten minutes later, Captain John E. Hurt, an F-18 pilot serving as aerial observer, launched in an Army 0H-58 Kiowa helicopter in support of the operation. At 1547, the company encountered a roadblock built around a number of buses. The LAVs began to take small arms fire from the roadblock and the adjoining hill. Only after the volume of fire increased did the crew members realize what was happening; the noise of their vehicles and the dampening effect of their helmets made it nearly impossible to hear anything else. To them it sounded like someone beating on a tin can far away. At that point, Captain Gaskins ordered his lead elements to assault the roadblock. Led by two LAVs rolling side by side at speed and firing their main guns, the assault was over almost as soon as it began. Shocked by the light armored attack, the owners of the buses, which belonged to private companies, took fright and drove away.[88]

While the LAVs dealt with the roadblock, Captain Hurt proceeded to reconnoiter in the vicinity of the objective, which was a block-long compound of four large buildings surrounded by a six foot concrete wall studded with firing ports. With the objective in sight, Hurt began to take fire. "You could see the tracers come up. In a chopper, you don't hear the guns; but you [do] hear the crack of the bullets coming by your aircraft. It sounds just like when you pull butts at the rifle range." Impressed by the volume of fire, and worried that the small Marine task force would be outnumbered and outgunned, Captain Hurt requested close air support for the impending assault from Marine Forces Panama, which decided that the risk of collateral damage was small and passed the request on to JTF South for action. Despite his initial opposition to running a close air mission in a built-up area, General Stiner himself approved the request. Within the hour, three aircraft were on station, one OA-37 Dragonfly to mark the target and two Virginia National Guard A-7 Corsairs to attack it. Captain Hurt controlled the mission. After the Dragonfly had fired a white phosphorous rocket into the compound, the A-7's stitched the target with 20mm cannon fire without damaging any buildings outside its walls.[89]

Seconds after the last cannon round struck the compound, Captain Hurt cleared the company to assault from the position it had assumed some 2,000 meters from the objective. Literally before the smoke had cleared, the LAVs passed through the main entrance of the compound, firing their main guns. The remnants of the PDF garrison, reported to have been 265 strong on 10 December, fled in disarray. The only opposition came from a few snipers who continued to fire while FAST Marines and LAI scouts cleared the buildings, which caught fire. Black smoke billowed through the compound. Within 10 minutes it was secure. The Marines seized a quantity of weapons and documents, and then returned to Arraijan with their booty since they were not strong enough to maintain a presence in La Chorrera and accomplish their original mission.[90]

PDF Prisoners

The Marine command's orders for D-Day plus 1, 21 December, were to avoid decisive engagements, since the focus of effort was going to other units. Nevertheless, there was pressing work. By the

Photo by Sgt Robert C. Jenks

After Marine LAVs entered the La Chorrera compound, LAI scouts prepared to assault and clear the reported 265-man-strong PDF garrison.

As thick black smoke billows from the compound's main building, Marine scouts wait for the final go-head to assault. Overcoming sporadic sniper fire, they secured the building within minutes.

Photo by Sgt Robert C. Jenks

early evening of 20 December, the command had learned that a crowd of 200 to 300 Panamanians was looting and burning buildings in Arraijan. It was for that reason that the command's missions for 21 December were to stabilize Arraijan, capture any PDF and Dignity Battalion personnel who were still at large, and expand its sphere of control to the nearby towns of Nuevo Arraijan and Vista Allegre.

The command looked first to Company D, already deployed around Arraijan and able to field urban security patrols on short notice, and then to a provisional rifle platoon which arrived a day later. Late in the morning on the 21st, Marine Forces Panama received an unexpected windfall that helped to accomplish its new mission. A Panamanian lieutenant by the name of Lt Danilo Gallardo surrendered to Security Force Marines at the western roadblock near Arraijan, and said that his men were ready to surrender if promised good treatment. After receiving that assurance, Gallardo kept his word, and, over the next two days, helped to arrange the surrender of more than 400 Panamanian soldiers.[91]

Much the same scenario occurred in Veracruz.

Photo by Sgt Robert C. Jenks

A FAST company Marine escorts one of the many detainees seized during the first hours of Operation Just Cause.

Several of the more than 400 Panamanian soldiers who surrendered at Arraijan board school buses which would take them to the Marine holding facility at Rodman.

Photo by Sgt Robert C. Jenks

On 22 December, Company K Marines returned to the center of that town to deter snipers and looters, and soon accepted the surrender of 53 PDF soldiers. The Marines decided to use a number of the former soldiers to guide security patrols that the Marine command had ordered for Veracruz in addition to Arraijan. Since all of these guides had local ties, the theory was that they would feel a sense of responsibility for their hometowns. Company K also enlisted the services of the woman who had been elected mayor but denied office by the Noriega regime. She helped the Marines with the restoration of city services, including the distribution of food and water, which had been cut off since 20 December, an unintended side-effect of the Marine roadblocks.[92]

After hearing reports of a rally by 500 Dignity Battalion members in La Chorrera on the morning of 22 December, and being directed by JTF South to maintain a presence in the city, Colonel Richardson and his staff made preparations to deploy their newly arrived reinforcements to La Chorrera. The reinforcements were the long-promised soldiers of the 2d Battalion, 27th Regiment, 7th Infantry Division (Light). To prepare for the deployment, Colonel Richardson and Lieutenant Colonel Franks reconnoitered by air. As they flew over the area by helicopter, they scattered surrender leaflets, which worked far better than anyone had imagined. Even before the deployment of the Army battalion, 75 PDF soldiers decided to surrender in La Chorrerra, and the LAI Marines had to make another trip to escort them to the closest detention center.[93]

The final event of the day was the insertion of the first wave of the 2d Battalion, 27th Infantry into La Chorrera by helicopter, which was completed during the early morning hours of 23 December. The battalion immediately swept through the city from southwest to northeast, driving a number of Panamanian soldiers towards the LAI and FAST Marines' positions screening the city to the east. During the days that followed, the battalion, like its Marine counterparts, spent much of its time capturing and guarding PDF soldiers who had evaded the first sweep.[94]

The detainees absorbed a great deal of time and effort. By the time Just Cause ended a few days later, the Marine command, numbering about four percent of the American forces, had captured some 1,320 Defense Force and Dignity Battalion members, about 25 percent of the total number detained by the American joint task force. The routine was for the unit that captured a detainee to send him to a forward collection point, such as the one in Arraijan. Literally hundreds of detainees poured into that site, which after 22 December was under the control of the U.S. Army's 534th Military Police Company, 16th Military Police Brigade. Working with the Marines, the military police searched, tagged, and processed the detainees before evacuating them to the Marine holding facility at Rodman.[95]

Photo by Sgt Robert C. Jenks

An exhausted Marine breaks for chow after long hours of confronting, detaining, and then processing large numbers of looters in the port of Vaca Monte.

At the facility, which was little more than a softball field with a few tents and portable toilets surrounded by barbed wire and guard towers, there was further processing. Between 20 December and 5 January, more than 2,000 Panamanians passed through the facility, and were checked against the data bases which the command's intelligence section had put together in the months before the operation. With senior PDF members helping to maintain order while armed sailors with dogs manned the perimeter, the camp ran smoothly. There were no recorded instances of mistreatment or attempts to escape, either from Rodman or from Arraijan.

In due course, the Navy and Marine Corps detainees were transferred to the JTF camp at Empire Range near the Canal for disposition. With few exceptions, the detainees were released by mid-January, the process of sorting the good from the bad having been completed. Apart from the individuals actually charged with crimes and held for trial, the task was to determine who was fit to return to public service and who was not. It was a subtle undertaking, complicated by the fact that the JTF's role was now to advise and support the new Panamanian government; American liaison officers could only make recommendations about the employment of any given individual, or about personnel policies in general. Unfortunately, some of the newly released detainees were PDF members believed to be corrupt, who somehow joined the post-PDF police force, much to the dismay of Marines like Colonel Richardson. In some cases, Marines detained the same man more than once. However, there were also many cases of PDF members who, under the new regime, did their best to rebuild the country.

More New Responsibilities

By 23 December, it was clear that there was little threat to the Arraijan Tank Farm. Duty for Security Force Marines in the storage facility was uneventful and their commander, Major Neller, approached Colonel Richardson about the possibility of detailing some of his Marines for more active duty. At that time, "more active duty" meant the impending operation against Vaca Monte, a nearby seaport. Marine Forces Panama had received reports that Noriega's 100-strong antiterrorist unit maintained a presence at a small base in the town.[96]

Neller got more than he bargained for. Colonel Richardson not only agreed to detach a number of Security Force Marines for the operation, but also put Neller in command. Richardson gave Neller two hours to prepare to move his new task force, codenamed Bulldog, composed of Company B, 2d Battalion, 27th Infantry; the LAI company, still reinforced with a detachment of FAST Marines; and the detachment of Security Force Marines. The Marine units began a motor march to the west, while the Army company, which started from La Chorrera, moved east to meet them at the intersection of the Pan-American Highway and the road to Vaca Monte. They rendezvoused at 1530.

With Captain Hurt once again reconnoitering to the front from the air, the units coordinated their plans while on the move. The coordination served its purpose, as it prevented the task force from inadvertently firing on U.S. Navy SEALs in high-speed boats heading north towards the objective. While the Navy screened the seaward flank, the task force entered the port's gateway, and encountered a crowd of looters, many of whom were trying to drive away on heavily loaded trucks. The Marines stopped the trucks, and threw the contraband on the ground. As word of the Marines' arrival spread, other looters and civilians began to stream towards the gateway, by now secured by Company B . All told, the task force detained some 400 people for Marine counterintelligence teams to process. The group included a number of foreign seamen, among them Cubans, Germans, Peruvians, and Nicaraguans, all from ships docked at Vaca Monte. By midnight all of the foreigners had been released, and the number of detainees had fallen to 100 Panamanians, 13 of them regular Panamanian soldiers assigned to Vaca Monte, who were held for further processing.

Confronting the looters turned out to be the highlight of the operation. While they occupied the attention of the lead elements of the task force, other elements searched for signs of Noriega's elite force, including supposed weapons caches, but found little apart from 120 kilograms of TNT. The antiterrorist unit had left the area well before the operation. When Colonel Richardson heard the news, he regretted that he had been unable to do more to neutralize a dangerous part of the Panamanian Defense Force.[97]

The other significant event of the day occurred when Marine Forces Panama was placed under the operational control of the 7th Infantry Division at

Photo by Sgt Robert C. Jenks

Vehicles and people wait to be searched at one of several checkpoints set up by Marines throughout their operational area.

U.S. Army military police attached to the Marine Command conduct a vehicle search.

Photo by Sgt Robert C. Jenks

1630 and, until the beginning of February, ceased to be a more or less independent component of JTF South. While the attachment was, at least in theory, a blow to Marine doctrine and pride, the practical consequences were limited, since the Marine Command had never been a true Marine air-ground task force and had already accom plished most of its assigned missions.[98]

Christmas Eve

The operation against Vaca Monte was the last major action which carried with it the threat of hostilities on a large scale. By Christmas Eve, Marine commanders realized that the fight was all but over, especially after Noriega sought refuge in the Papal Nunciatura. When the news spread, crowds of Panamanians celebrated in the streets. In Veracruz, they produced a large portrait of the legitimately elected president, Guillermo Endara, and hung it at the local PDF station. The hanging of the portrait of Endara marked the end of hostilities in Veracruz and the surrounding area.[99]

Christmas Day and After

During the days that followed, the Marines slipped into a routine of search and seizure operations, and conventional patrolling. The searches were still mainly for PDF cadre or weapons. There were, basically, two types of searches, those at roadblocks and those of property. Working the roadblocks was frustrating. After 0600 on 25 December, JTF South allowed vehicles to pass through the blocks, but only under the condition that they be searched. The intent was to search vehicles quickly but thoroughly. It did not work that way. The buses that stopped at the roadblocks were filled to overflowing with the poor, their domestic animals, and their possessions. The initial reaction among Marines was that it was simply impossible to accomplish the mission. However, over time, experts emerged who could do the work. There was, for example, a Security Force Marine who developed an unerring sense for PDF soldiers, whom he could distinguish from ordinary citizens, and detain.[100]

The property searches were largely conducted by FAST Marines supported by LAI Marines, whose duty in Vaca Monte had ended by Christmas Day, freeing them for other tasks. Typically, the LAVs secured the perimeter, while FAST cleared the objective. All told, they cleared approximately 40 multistory, multiroom buildings under the leadership of First Lieutenant Killackey, who had returned to Panama for a second tour. The targets even included a radio station and possible tunnel complex in the town of Nuevo Emperador. During these operations, LAI and FAST Marines captured a number of Noriega affiliates, all without causing any civilian casualties.[101]

But perhaps the most important capture happened almost by chance. It was during a vehicle recovery operation on Christmas Day that a detachment of support group Marines captured local strongman Rigoberto Paredes while passing through Nuevo Arraijan. Paredes, who was high on JTF South's wanted list, had eluded U.S. forces for five days. When the detachment appeared on the scene, local citizens were literally chasing Paredes down the street. Being captured saved his life. Even after the capture, the citizens pleaded with the detachment commander, Captain William Aitken, to kill him.[102]

Marine Reinforcements

While some Marines who had experienced months of tension, followed by the release of Just Cause, and then the letdown after combat, began to feel a sense of frustration, a fresh company of Marines, Company I, 3d Battalion, 6th Marines, was preparing to embark for Panama. The company's Marines had gone on their normal Christmas leave. On 27 December, the regimental commander asked the company commander if he could have his Marines ready to deploy by noon on 28 December. In an act of faith, Captain Blaine D. Vorgang assured Colonel Sutton that he could. There were a few tense hours. At 0800 on 28 December, he had only 72 Marines in his squadbays but, by noon, he had 241 Marines and was able to embark the company for Panama the same afternoon. When they landed, they brought the total number of Marines in Panama to 927, compared to approximately 21,000 Army, 3,500 Air Force, and 550 Navy personnel.[103]

On arrival, the company wasted little time. Early on the morning of 29 December, Colonel Richardson briefed the new unit on its mission, which he described as public safety and civil affairs. Later, Vorgang's Marines drew ammunition and equipment, and the next day, they deployed to Nuevo Arraijan, Nuevo Emperador, and Vista Allegre. They also replaced Security Force Marines on the roadblock west of Arraijan and detached one platoon to the force guarding the tank farm. Although their instructions were somewhat nebulous–establish a presence in the towns and keep the peace–and although the company was thin on the ground, given the size of its area of operations, all of the Marines turned to with enthusiasm, including a provisional rifle platoon of forward observers, engineers, mortarmen, machine gunners, and armorers. For these Marines new to Panama, it was different, interesting, and challenging work. Not only did they secure and patrol the towns, but they also built a small bridge in Nuevo Arraijan.[104]

At times, domestic, military, and political pressures combined in strange ways. On the night of 31 December, Company I Marines deployed to search a pig farm after receiving a tip that there was a Dignity Battalion arms cache on the property. (Marine Forces Panama had received reports that some Dignity Battalions planned to resist even after most of their comrades in the PDF had surrendered.) Shortly after midnight, in the first hours of the new year, the Marines took fire from the farm, ran through the rules of engagement, returned 17 rounds of fire, and wounded one Panamanian in the leg. Only after the firing had ceased did they learn that the "hostiles" were paid security guards; they had not known that American forces were in the area, and had been afraid that someone was trying to steal their pigs.[105]

The pattern was similar on 3 January, the day Noriega walked out of the Papal Nunciatura, past the two Marine LAVs still attached to Task Force Gator, and surrendered to members of the U.S.

Marines take to horseback to patrol the countryside in search of former PDF members and arms caches.

Photo by CWO-3 Charles W. Rowe, Jr.

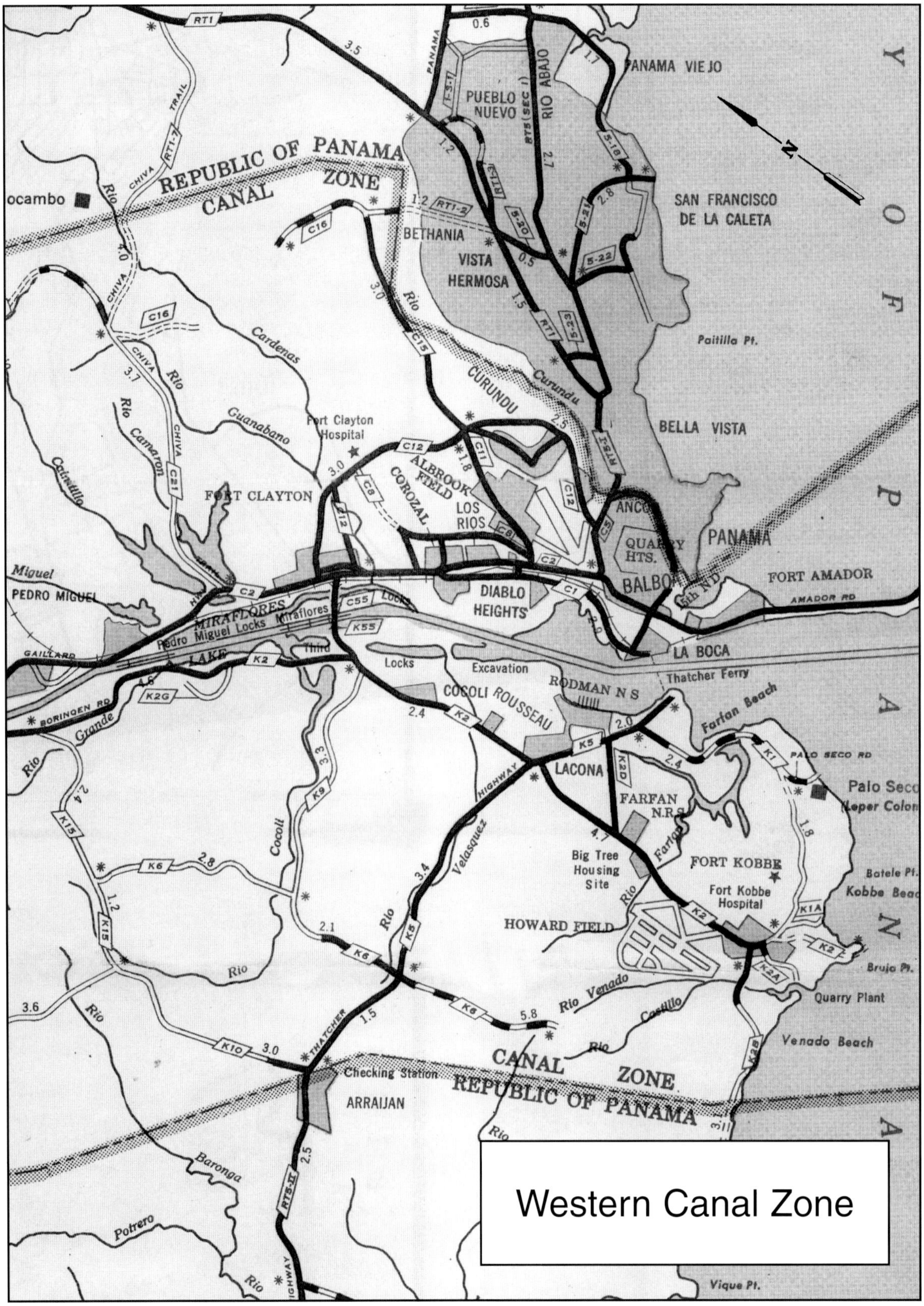

Western Canal Zone

Army's Delta Force. Against a backdrop of Panamanians beating pots and pans, their way of urging Noriega on his way to prison, Captain Vorgang received nearly simultaneous reports that two of his detachments, the Marines at the roadblock and a patrol in a rock quarry, were taking fire. There were also reports of a possible Panamanian attack on Vaca Monte, which he thought could be the kind of last-ditch stand predicted in some intelligence reports. It was impossible to determine over the radio what had happened, but Vorgang had an obligation to pass the report on to Marine Forces Panama, which deployed a reaction force of LAVs. Then the threats dissipated. But they had been real. The report from the roadblock grew out of a domestic incident in which a battered woman came to the Marines for assistance.When they went to investigate, someone started to shoot at them with a pistol. The Marines did not return fire. The attack at the quarry was also real enough, but it was simply harassing fire. Vorgang debriefed the patrol leader carefully, testing his story. His answer was, "Sir, when the ground next to you starts spurting up, and you see muzzle flashes, and you hear the crack over your head . . . you're receiving fire." Nevertheless, the Marines did not return fire because they did not have a clear target.[106]

During January, the operations of Task Force Cronin in Arraijan also reflected the changing mix of political and military affairs. Commanded by Captain R. Barry Cronin, a Marine Corps military police officer assigned to the command element of Marine Forces Panama, the task force was an ad hoc group of some 70 Marines and soldiers which had started as a contingency platoon. They first arrived in Arraijan on 22 December to help fill the vacuum left by the collapse of the Noriega regime, and started from scratch, with a hasty all-around defense and makeshift fortifications. Cronin spent the first night catnapping in a truck, pistol in hand, and awakened more than once to gunfire as his Marines traded fire with Dignity Battalion members. In the days that followed, they conducted patrols, intended primarily to prevent the PDF and the Dignity Battalions from organizing any further resistance. Then the focus shifted to helping to manage the collection point in Arraijan for detainees. When the flood of detainees ebbed, the task force focussed more and more on preventing crime and restoring order with the welcome assistance of Army Captain Steve Stanaland and his detachment of civil affairs specialists from the 96th Civil Affairs Brigade. There was food to distribute, a clinic to reopen, and a police force to organize. The task force installed the rightfully elected mayor who, as other Panamanian politicians, had been elected in May 1989 but denied office, and, alongside him, a police chief chosen by the townspeople. They helped him to recruit a police force which ran joint patrols with the Marines.[107]

By early January, city services had returned to the level where they were on 20 December. Over time, the Marines began to see their role not so much as governing the town as of protecting the new government and helping it to function. At first the population was somewhat standoffish, hesitant to work with the Marines. After Noriega was captured on 3 January, there was a turning point in Arraijan, as in other towns. The PDF having been defeated, it was now safe for Panamanians to help the Marines, and many did so willingly. They offered information on arms caches, pointed out former PDF and Dignity Battalion members for arrest, and in general helped to restore order. When the situation in Arraijan itself stabilized, Cronin sent mobile detachments out into the surrounding countryside, and repeated much the same process of removing the traces of the Noriega regime and strengthening its successor.[108]

Operation Promote Liberty

By the end of January, the transition from combat operations to nation-building appeared complete. Many elements of the Army's XVIII Airborne Corps had redeployed to the United States, JTF South having been disbanded on 11 January. When the 7th Infantry Division left Panama in early February, Marine Forces Panama once again became a component of JTF Panama, as it had been before 20 December. But there was a new title for the new phase of operations. In place of Operation Just Cause, there was now Operation Promote Liberty. Its purpose was "to promote democracy, promote Panamanian confidence in its democratically elected government, and support the nation-building process."[109]

During Promote Liberty, the Marine command's mission was to accomplish the JTF mission in its area of operations in addition to continuing to screen Howard and guard the two storage facilities. There was little activity in the tank farm or the ammunition supply point, only occasional intruders who appeared to be poachers. Even so, the Marines who stood duty in the fuel depot faced most of the same training requirements as had their predecessors, participating in rules of engagement exercises and spending a great deal of time on the rifle and pistol ranges. The command also experimented with methods to minimize the dangers of providing security at fixed sites, such as camouflaging sentry posts, using infra-red lights to communicate, and infiltrating reliefs through dense terrain.[110]

The absence of activity in the tank farm and ammunition depot allowed the Marine units to focus on other missions. Usually in concert with Panamanian police, they continued to patrol the towns of Veracruz and Arraijan, and to help to restore their municipal governments. During the period leading up to Promote Liberty, there were still platoon-sized units in the former PDF garrisons, or cuartels, of both towns. Toward the end of the operation, it was, for political reasons, necessary to let the new Panamanian National Police stand on its own. Accordingly, the Marine presence in each town was reduced to a few men, occasionally augmented by small detachments to help patrol or to provide replacements for observation and listening posts. Marine Reservists from the 4th Civil Affairs Group on two-week tours of active duty were among the augmentees. They coordinated relief efforts, screened candidates for government positions, documented claims against the U.S. Government, and liaised with local officials. Colonel Richardson recognized the Reservists' contribution, which demonstrated to him the need for organic civil affairs support in operations such as Just Cause and Promote Liberty.[111]

Throughout the period, the Marine command continued to gather tactical intelligence. Marine intelligence specialists in civilian clothes, using the cover of being on liberty, conducted surveillance in the vicinity of the cuartels Sentries maintained logbooks to note anything of interest, such as vehicles that drove by more than once. Members of the command's intelligence section travelled to the cuartels on a daily basis to debrief sentries, review logbooks, and update their own data bases. Other Spanish-speaking interrogator-translator and counterintelligence Marines circulated among the townspeople to the maximum extent possible, in some areas literally going from door to door to establish rapport and gather information openly. The primary, but not exclusive, focus of their efforts was to assess threats to Marine units. The

Several small Marine task forces, as this one, moved out into central and western Panama to conduct reconnaissance and assistance missions as Panama gradually returned to a democratic society.

Photo by Sgt Robert C. Jenks

Photo by Sgt Robert C. Jenks

During the height of the fighting two children in Vera Cruz took time to make a sign thanking the Marines of Company K, 3d Battalion, 6th Marines.

secondary purpose was to prepare packages of information for civil affairs operations. Both were extremely useful to commanders.[112]

During Promote Liberty, Marine elements went far beyond their originally assigned area of operations, and conducted reconnaissance and presence missions in central and western Panama. In early February, the LAI company deployed to the towns of Santiago and Sona, and, in late February, an infantry task force deployed to the Azuero Peninsula. During March, a completely motorized and mechanized task force under Marine Forces Panama's last commanding officer, Colonel James M. Hayes, launched Operation Texas Rangers, a reconnaissance in force in and around the towns of David, Volcan, and Rio Sereno in the province of Chirique. Built around Captain Richard R. Huizenga's Company C, 2d LAI Battalion, the task force covered 1,500 kilometers and made five tactical overnight stops. Its accomplishments included the capture of holdouts from the Noriega regime, in one case within meters of the Costa Rican border. Earlier, the task force uncovered a cache of falsified passports and large amounts of American currency. During the operation, the Marines enjoyed the support of courageous Panamanians like the mayor of Rio Sereno, Mario Lascano, who rode along with the LAVs, pointing out targets and signing search warrants. In addition to search and seizure operations, the Marines did what civil affairs work they could. For example, they opened field expedient roadside clinics and provided medical care to the local citizenry while they were in the area.[113]

In late March, Marines replaced Army troops at the old Panamanian Defense Force base at Rio Hato, a town located some 80 kilometers west of the Canal. The mission was to maintain a presence and a forward refueling site. The Marine command took advantage of the opportunity, and used Rio Hato to support its LAI operations in the west. For First Lieutenant Brian K. Foster, a Company C platoon commander, the deployment was a tremendous opportunity. Although an unusual responsibility for an officer of his rank, he was placed in command of a task force which comprised roughly 80 Marines and sailors built around a rifle platoon and an LAI platoon. He learned quickly that there was no one to rely upon except himself. For 51 days he was, in effect, the mayor of Rio Hato and the surrounding countryside.[114]

Foster used his small task force well, planning and executing a judicious mixture of deterrent patrols and nation-building operations. At first the Marines and sailors worked hard simply to maintain a presence and establish rapport with their neighbors in the provincial towns, while obeying orders not to become involved with any of the young women in them, which Foster later described as the greatest potential danger to the success of his mission. Over time, the Marines won the support of the local citizenry. They discovered, for example, that they could use their tanquitos to accomplish their peacetime mission. In the words of the Marine command's operations officer, Lieutenant Colonel Robert J. Fawcett, it was "the light armored vehicle people-magnet."[115] They attracted crowds in remote areas simply by appearing. The Marines would then distribute candy to the children, and take the opportunity to make contact with curious townspeople. In a short time, they became more comfortable with the Marine presence, and did what they could to help, offering information and hospitality. That in turn made it possible for the Marines to help the National Police in another way; they ferried a number of policemen around the countryside in the tanquitos and then introduced them to contacts that they had made earlier alongside the armored vehicles.[116]

Apart from other good works, like repairing roads, bridges, and other facilities when possible, the Marines helped to reduce the danger from unexploded ordnance. After learning that a number of children had been wounded by unexploded ordnance, the Marines conducted a series of classes in local schools, describing the dangers and telling the children what to do if they found ordnance. The program was a success and proved to be yet another way to get close to the Panamanian people.[117]

In late April, political considerations dictated a curtailment of routine patrols in the outlying regions, and the LAI platoon was withdrawn from Rio Hato. The rifle platoon remained, but limited its activities to training and providing security for the base. At the same time, the Marine command conducted two small intelligence-gathering operations, sending motorized units to Arenas on the western side of the Azuero peninsula and to the mountainous region of Coclecito, 130 kilometers west of the capital.

By May, the situation was such that JTF Panama decided that it was no longer appropriate for U.S. forces to operate at any distance from the Canal, and plans were made to redeploy the Marine command to the United States. By 1 June all detachments still deployed in the countryside had withdrawn to the Canal Zone and the pack-out began. On 13 June, all of the elements of the command, with the exception of Security Force Marines, embarked on board the *Trenton* and, five days later, the last Marine walked down the gangway at Morehead City, North Carolina. Marine Forces Panama was deactivated on 23 June 1990.[118]

The prevailing attitude among the Marines was that it was time to leave. They were proud of their accomplishments, and their morale was high. Lieutenant Colonel Fawcett spoke for many when he said: "Wait a second. You tell me to take the magazine out of my weapon and put it back in my pouch. You don't need any more rifle carriers. It's time for us to go home."[119]

Notes

1. For additional background see: Allan R. Millett, *Semper Fidelis, The History of the United States Marine Corps* (New York: The Free Press, 1980), pp. 165-167; Col Robert D. Heinl, Jr., *Soldiers of the Sea–The United States Marine Corps, 1775-1962* (Annapolis: U. S. Naval Institute, 1962), p. 148; and Capt J. H. Reynolds, "The Marines in Panama" (ms, MCHC, Washington, D.C., 1975).

2. Capt J. H. Reynolds, "The Marines in Panama" (ms, MCHC, Washington, D.C., 1975), pp. 10-12.

3. Maj Alfred F. Clarkson intvw, 27-28Mar90, pp. 4-5 (Oral History Collection, MCHC, Washington, D.C.), hereafter Clarkson intvw; and LtCol James B. Thomas intvw, 29May90, pp. 12-13 (Oral History Collection, MCHC, Washington, D.C.), hereafter Thomas intvw.

4. Clarkson intvw, pp. 6-7; and LtGen Ernest T. Cook intvw, 29May90, pp. 2-5 (Oral History Collection, MCHC, Washington, D.C.), hereafter Cook intvw.

5. Cook intvw, p. 2; Clarkson intvw, p. 6; and Jeremiah O'Leary, "White House says Marines to project power in Panama," *The Washington Times*, 4Apr88, p. 9.

6. COMarForPM, Sitrep 1, 072302ZApr88 ("Just Cause" File, MCHC, Washington, D.C.).

7. Maj Joseph P. Valore intvw, 29Mar90, pp. 4-7 (Oral History Collection, MCHC, Washington, D.C.), hereafter Valore intvw.

8. Ibid., pp. 9-10; and Maj William J. Philbin intvw with author, 23Jun94.

9. Clarkson intvw, p. 61.

10. Valore intvw, pp. 16-20.

11. Ibid., p. 20.

12. Cook intvw, pp. 12-13.

13. Valore intvw, p. 23.

14. Clarkson intvw, pp. 18-19.

15. Valore intvw, pp. 25-30.

16. Valore intvw, p. 31; and Clarkson intvw, pp. 19, 26.

17. Valore intvw, pp. 32-35; and Clarkson intvw, pp. 21, 24.

18. Ibid., p. 25.

19. Ibid., p. 31.

20. Ibid., p. 27; and Valore intvw, p. 51.

21. Clarkson intvw, p. 34; and Thomas intvw, pp. 23-24.

22. Clarkson intvw, p. 35; and COMarForPM, Sitrep 2, 152300ZApr88 ("Just Cause" File, MCHC, Washington, D.C.).

23. Clarkson intvw, p. 23.

24. Ibid., p. 36. See also: William Branigin, "Marines in Firefight in Panama," The Washington Post, 14Apr88, p. A-1, and William Branigin, "U.S. Marine Describes 'Intruders'; Doubts Persist on Panama Incident," *The Washington Post*, 16Apr88, p. A-18.

25. Valore intvw, p. 46.

26. Clarkson intvw, p. 37.

27. COMarForPM msg to LtGen Cook, 141232ZFeb90 ("Just Cause" File, MCHC, Washington, D.C.). See also: Maj Charles A. Christburg intvw, 25Apr90, p. 33 (Oral History Collection, MCHC, Washington, D.C.), hereafter Christburg intvw; and Col Charles E. Richardson intvw, 23Apr90, p. 48 (Oral History Collection, MCHC, Washington, D.C.), hereafter Richardson intvw.

28. COMarForPM, Sitrep 43, 270500ZJan89 ("Just Cause" File, MCHC, Washington, D.C.).

29. MarForPM ComdC, 6Apr88-31Dec88 ("Just Cause" File, MCHC, Washington, D.C.), p. 6.

30. Clarkson intvw, p. 44; Valore intvw, p. 60; and LtCol Gene A. Tromly intvw, 13Apr90, p. 44 (Oral History Collection, MCHC, Washington, D.C.), hereafter Tromly intvw.

31. Clarkson intvw, p. 69; Valore intvw, pp. 54, 76.

32. Clarkson intvw, pp. 47-48; COMarForPM, Sitrep 16, 220900ZJul88 ("Just Cause" File, MCHC, Washington, D.C.).

33. Clarkson intvw, pp. 47-48.

34. Tromly intvw, pp. 43-44.

35. Clarkson intvw, p. 56.

36. Ibid., p. 62; and LtCol William K. Jones, Jr., intvw, 17Apr90, pp. 7-8 (Oral History Collection, MCHC, Washington, D.C.), hereafter Jones intvw.

37. Clarkson intvw, p. 64; LtCol Martin R. Berndt intvw, 26Apr90, pp. 24-25 (Oral History Collection, MCHC, Washington, D.C.), hereafter Berndt intvw; and USCinCSo msg to JCS, 141831ZDec88 ("Just

Cause" File, MCHC, Washington, D.C.).

38. Jones intvw, p. 18.

39. LtCol Sands A. Robnick intvw, 24Apr90, p. 19 (Oral History Collection, MCHC, Washington, D.C.), hereafter Robnick intvw; and COMarForPM msg to LtGen Cook, 292030ZNov88 ("Just Cause" File, MCHC, Washington, D.C.).

40. Col Thomas W. Roberts intvw, 30Mar90, pp. 54-55 (Oral History Collection, MCHC, Washington, D.C.), hereafter Roberts intvw; and Robnick intvw, pp. 22-23.

41. Clarkson intvw, p. 69.

42. Ibid., p. 72; and Berndt intvw, p. 34.

43. Capt Walter L. Miller intvw, 26Apr90, pp. 1-2 (Oral History Collection, MCHC, Washington, D.C.), hereafter Miller intvw.

44. Robnick intvw, pp. 1-2; and COMarForPM Sitrep 66, 071900ZJul90 ("Just Cause" File, MCHC, Washington, D.C.)

45. Capt Edward J. Moss intvw, 25Apr90, p. 6 (Oral History Collection, MCHC, Washington, D.C.), hereafter Moss intvw.

46. Capt Scott E. Nisely, "Frequent Firing Helps," *Marine Corps Gazette*, Jul93, p. 42.

47. Clarkson intvw, p. 74; and Maj Robert B. Neller intvw, 17May90, pp. 34-35 (Oral History Collection, MCHC, Washington, D.C.), hereafter Neller intvw.

48. COMarForPM Sitrep 72, 180800ZAug89 ("Just Cause" File, MCHC, Washington, D.C.).

49. COMarForPM Sitrep Collection ("Just Cause" File, MCHC, Washington, D.C.).

50. See USCinCSo msg to JCS, 141731ZDec88 ("Just Cause" File, MCHC, Washington, D.C.).

51. For an overview of LAV employment in Panama, see Col Robert P. Mauskapf and Maj Earl W. Powers et al., "LAVs in Action," Marine Corps Gazette, Sep90, pp. 50-59, hereafter Mauskapf and Powers, "LAVs." See also Cook intvw, p. 1, for mention of General Woerner's attitude on LAVs.

52. COMarForPM msg to LtGen Cook, 290030ZMay89 ("Just Cause" File, MCHC, Washington, D.C.).

53. Mauskapf and Powers, "LAVs," p. 51.

54. Ibid., p. 52.

55. Maj Robert B. Neller, "Marines in Panama: 1988-1990," p. 15 (ms, MCHC, Washington, D.C., 1991), hereafter Neller, "Marines in Panama."

56. Mauskapf and Powers, "LAVs," pp. 52-53.

57. Ibid., p. 52.

58. Neller intvw, pp. 21-23; and First Lieutenant Robert F. Killackey intvw, 30May90, pp. 12-14 (Oral History Collection, MCHC, Washington, D.C.), hereafter Killackey intvw.

59. Ibid., pp. 15-17.

60. See COMarForPM Oplan 5-89 (Ghostbuster), 071600RNov89 ("Just Cause" File, MCHC, Washington, D.C.). For a general description of the planning phase, see LtGen Edward M. Flanagan, Jr., USA (Ret), Battle for Panama (Washington: Brassey's, 1993), pp. 32-45, hereafter Flanagan, Battle for Panama.

61. Ibid., p. 70. See also: LtCol Michael J. Franks intvw, 23Apr90, pp. 13-15 (Oral History Collection, MCHC, Washington, D.C.), hereafter Franks intvw; Cook intvw, p. 7; and Richardson intvw, pp. 6, 22.

62. Mauskapf and Powers, "LAVs," p. 53.

63. Ibid., p. 54.

64. Ibid., p. 54.

65. Ibid., p. 54.

66. 1stLt Kenneth M. DeTreux, "Infantry in Low-Intensity Conflict: Observations from Operation Just Cause" *Marine Corps Gazette*, Sep90, p. 65, hereafter DeTreaux, "Just Cause"; and Richardson intvw, p. 32.

67. DeTreux, "Just Cause," p. 65; Killackey intvw, p. 17; and Members of 1st Light Armored Infantry Battalion intvw, 28Mar90, p. 15 (Oral History Collection, MCHC, Washington, D.C.), hereafter LAI intvw.

68. Richardson intvw, p. 11.

69. 2dLt Thomas H. Goessman intvw, 17May90, p. 17 (Oral History Collection, MCHC, Washington, D.C.).

70. Maj Robert C. Griffin, "Investigation to Inquire into the Circumstances Surrounding the Death of First Lieutenant Robert Paz, USMC, on 16 December 1989," 7Jan90 ("Just Cause" File, MCHC, Washington, D.C.).

71. Flanagan, Battle for Panama, pp. 51-52.

72. Ibid., p. 71; and Richardson intvw, p. 31.

73. Maj James W. Schindler intvw, 16May90, p. 10 (Oral History Collection, MCHC, Washington, D.C.); Neller, "Marines in Panama," pp. 18-19; and CWO-3 C. W. Rowe, MarForPM Press Release 03-90, n.d. ("Just Cause" File, MCHC, Washington, D.C.).

74. CWO-3 C. W. Rowe, MarForPM Press Release 20-90, n.d. ("Just Cause" File, MCHC, Washington, D.C.).

75. CWO-3 C. W. Rowe, MarForPM Press Release 08-90, n.d. ("Just Cause" File, MCHC, Washington, D.C.); LAI intvw, p. 30; and Maj Bron N. Madrigan,

"The Attack on the DNTT Station," *Marine Corps Gazette*, Apr90, pp. 9-10.
76. Franks intvw, pp. 17-18; LAI intvw, pp. 18-20.
77. Franks intvw, pp. 16-17; and Mauskapf and Powers, "LAVs," p. 57.
78. Franks intvw, p. 18; and Mauskapf and Powers, "LAVs," p. 57.
79. Ibid., p. 57; Franks intvw, p. 19.
80. MarForPM, "Sequence of Events, OpOrd 5-89 "Ghostbuster"/"Just Cause," 20-23Dec89, p. 4 ("Just Cause" File, MCHC, Washington, D.C.), hereafter "Sequence of Events." See also MarForPM ComdC, 1Jul89-31Dec89.
81. "Sequence of Events," p. 4; and Capt William S. Aitken, "Combat Service Support During Operation Just Cause," *Marine Corps Gazette*, Dec90, pp. 24-25.
82. Mauskapf and Powers, "LAVs," p. 58.
83. CWO-3 C. W. Rowe, MarForPM Press Release 01-90, n.d. ("Just Cause" File, MCHC, Washington, D.C.); Maj Bron N. Madrigan, "How I Spent the Holidays," p. 43 (ms, MCHC, Washington, D.C., 1990), hereafter Madrigan, "Holidays"; and "Sequence of Events," pp. 2-3.
84. Madrigan, "Holidays," p. 43.
85. CWO-3 C. W. Rowe, MarForPM Press Release 09-90, n.d. ("Just Cause" File, MCHC, Washington, D.C.); "Sequence of Events," p. 2; and WO Scott Whitney intvw, 17May90, pp. 6-10 (Oral History Collection, MCHC, Washington, D.C.), hereafter Whitney intvw.
86. Capt Donald D. Cline, Untitled Timeline of Activities for Company K in "Just Cause," 15Jan90 ("Just Cause" File, MCHC, Washington, D.C.).
87. Mauskapf and Powers, "LAVs," pp. 57-58; and "Sequence of Events," pp. 3-5.
88. CWO-3 C. W. Rowe, MarForPM Press Release 02-90, n.d. ("Just Cause" File, MCHC, Washington, D.C.); and Mauskapf and Powers, "LAVs," pp. 57-58.
89. Madrigan, "Holidays," pp. 9-10.
90. Mauskapf and Powers, "LAVs," p. 58; and COMarForPM msg to CG, FMFLant, 281232ZDec89 ("Just Cause" File, MCHC, Washington, D.C.).
91. "Sequence of Events," p. 5; Madrigan, "Holidays," p. 11; and COMarForPM msg to LtGen Cook, 281232ZDec89 ("Just Cause" File, MCHC, Washington, D.C.).
92. Maj Donald D. Cline intvw, 25Mar90, p. 27 (Oral History Collection, MCHC, Washington, D.C.); and "Sequence of Events," p. 8.
93. Ibid., pp. 7-8.
94. Ibid., pp. 8-9; and COMarForPM msg to LtGen Cook, 281232ZDec89 ("Just Cause" File, MCHC, Washington, D.C.).
95. CWO-3 C. W. Rowe, MarForPM Press Release 18-90, n.d. ("Just Cause" File, MCHC, Washington, D.C.); Flanagan, Battle for Panama, p. 211; and Capt R. Barry Cronin, "Military Police Operations in Panama," *Marine Corps Gazette*, Sep90, pp. 67-68.
96. "Sequence of Events," p. 9; and Neller intvw, pp. 47-48.
97. Richardson intvw, p. 24.
98. Ibid., pp. 28-29; and "Sequence of Events," p. 9.
99. Cline intvw, p. 39.
100. Madrigan, "Holidays," p. 27; "Sequence of Events," p. 13.
101. Killackey intvw, pp. 26-27; and "Sequence of Events," p. 12.
102. Ibid., p. 12; and Madrigan, "Holidays," p. 25.
103. Capt Blaine D. Vorgang intvw, 26Apr90, p. 8 (Oral History Collection, MCHC, Washington, D.C.), hereafter Vorgang intvw.
104. Ibid., pp. 11-19.
105. Ibid., pp. 25-26.
106. Ibid., pp. 27-31.
107. Otto Kreisler, "Marines, Soldiers Help Panamanians Pass Civics 101," *San Diego Union*, 2Jan90, pp. A-1, A-18.
108. Ibid.
109. Col James M. Hayes et al., "Panama Promote Liberty After Action Report," Marine Corps Gazette, Sep 90, p. 59, hereafter Hayes, "Promote Liberty."
110. Ibid., pp. 60-61; and LtCol Robert J. Fawcett intvw, 17May90, pp. 21-28, 30-31 (Oral History Collection, MCHC, Washington, D.C.), hereafter Fawcett intvw; and MarForPM, Promote Liberty AAR, 17May90 ("Just Cause" File, MCHC, Washington, D.C.), hereafter MarForPM, Promote Liberty AAR.
111. COMarForPM msg to LtGen Cook, 292359ZJan90 ("Just Cause" File, MCHC, Washington, D.C.); and Richardson intvw, p. 38.
112. Hayes, "Promote Liberty," p. 62; and Richardson intvw, p. 13. See also Whitney intvw.
113. Hayes, "Promote Liberty," pp. 58-59; and MarForPM ComdC,
1Jan90-30Jun90, pp. 9-10.
114. 1stLt Brian K. Foster intvw, 16May90, pp. 10-20 (Oral History Collection, MCHC, Washington, D.C.), hereafter Foster intvw.

115. Hayes, "Promote Liberty," p. 61.
116. Foster intvw, p. 20.
117. Ibid., p. 27.
118. USCinCSo msg to CMC, 151615ZJun90 ("Just Cause" File, MCHC, Washington, D.C.); CG, FMFLant 231301ZJun90 ("Just Cause" File, MCHC, Washington, D.C.).
119. Fawcett intvw, p. 411 See also: COMarForPM msg to LtGen Cook, 281232ZDec89 ("Just Cause" File, MCHC, Washington, D.C.).

Appendix A
Troop List

Marine Forces Panama

Command Element

Marine Corps Security Force Company, Panama

Platoons detached from First Fleet Anti-Terrorist Security Force Company,

Marine Corps Security Force Battalion, Atlantic

Detachments, Brigade Service Support Group-6

Ground Combat Element

Company I, 3d Battalion, 4th Marines (6Apr88-10Jul88)

Company L, 3d Battalion, 4th Marines (6Jul88-5Oct88)

Company M, 3d Battalion, 8th Marines (29Sep88-13Jan89)

Company C, 1st Battalion, 6th Marines (3Jan89-18Apr89)

Company B, 1st Battalion, 6th Marines (10Apr89-7Jul89)

Company L, 3d Battalion, 6th Marines (7Jul89-10Oct89 and 9Apr89-13Jun90)

Company K, 3d Battalion, 6th Marines (8Oct89-1Feb90)

Company I, 3d Battalion, 6th Marines (28Dec89-8Apr90)

Company A, 2d LAI Battalion (12May89-6Aug89)

Company B, 2d LAI Battalion (7Aug89-6Nov89)

Company D, 2d LAI Battalion (26Oct89-14Feb90)

Company C, 2d LAI Battalion (13Feb90-13Jun90)

Appendix B
Command and Staff, Marine Forces Panama

Commanding Officer

Col William J. Conley–6Apr88-5Oct88
Col Thomas W. Roberts–5Oct88-23Mar89
Col James J. Doyle, Jr.–24Mar89-16Sep89
Col Charles E. Richardson–17Sep89-5Mar90
Col James M. Hayes–6Mar90-23Jun90

Executive Officer

Col James M. Puckett–1Jun88-8Jul88
LtCol Walter W. Shallcross–9Jul88-16Sep88
LtCol Martin R. Berndt–17Sep88-12May89
LtCol Sands A. Robnic–13Mar89-24May89
LtCol Larry G. Clarke–25May89-12Nov89
LtCol Slade A. Brewer–13Nov89-15Apr90
LtCol James Q. Butler, Jr.–16Apr90-23Jun90

S-3

Col Donald F .Anderson–6Apr88-30May88
LtCol Edward J. Ball III–31May88-28Jul88
Maj Bruce A. Denault–29Jul88-23Aug88
Maj Jack K. Sparks, Jr.–24Aug88-30Aug88
LtCol William K. Jones, Jr.–31Aug88-26Nov88
Maj Alfred E. Clarkson, Jr.–27Nov88-27Feb89
LtCol John D. Winchester–28Feb89-18Jun89
LtCol Edward J. Robeson IV–19Jun89-6Sep89
LtCol Michael J. Franks–7Sep89-23Feb90
LtCol Robert J. Fawcett–24Feb90-23Jun90

Commanding Officer, Marine Corps Security Force Company

Maj Eddie A. Keith–6Apr88-8Jul88
1stLt Glen Alberto–9Jul88-4Aug88
Maj Robert B. Nelle–5Aug89-23Jun90

Company Commanders

Company I, 3/4: Capt Joseph P Valore, 6Apr88-10Jul88
Company L, 3/4: Capt Dale L. Davidson, 6Jul88-5Oct88
Company M, 3/8: Capt Byron G. Richardson, 29Sep88-13Jan89
Company C, 1/6: Capt William L. Miller, Jr., 3Jan89-18Apr89
Company B, 1/6: Capt Edward J. Moss, 10Apr89-7Jul89
Company L, 3/6: Capt Daniel E. Liddell, 7Jul89-10Oct89
Company K, 3/6: Capt Donald D. Kline, 8Oct89-1Feb90
Company I, 3/6: Capt Blaine D. Vorgang, 28Dec89-8Apr90
Company L, 3/6: Capt William L. Blair II, 9Apr90-13Jun90

Light Amphibious Infantry Company Commanders

Company A: Capt Stephen J. Linder, 12May89-6Aug89
Company B: Capt John S. Dunn, 7Aug89-6Nov89
Company D: Capt Gerald H. Gaskins, 26Oct89-14Feb90
Company C: Capt Richard R. Huizenga, 13Feb90-13Jun90

Appendix C
Chronology

1988

February	Noriega indicted in U.S. by federal grand juries; unrest in Panama.
14 March	Platoon from Fleet Anti-Terrorist Security Team Company reinforces Marine Security Force Company (MCSF) Panama.
6 April	300 additional Marines, including a rifle company, arrive in Panama to further enhance the security of Rodman Naval Station.
12 April	First major firefight in Arraijan Tank Farm occurs between Marines and Panamanian troops.
25 April	Southern Command activates Area of Operation Pacific West, giving Marine Forces Panama authority over all Marines in Panama, as well as Army military police and infantry units, to provide security for Rodman, Howard, and surrounding American facilities.
31 October	Major firefight occurs between Marines and Panamanians in the tank farm.

1989

7 May	Noriega supporters lose national elections; further unrest and challenges to U.S. forces, especially to their freedom of movement.
12 May	Additional U.S. reinforcements, including a Marine light amphibious infantry company (LAI), arrive in Panama.
22 July	The Joint Chiefs of Staff issue National Security Directive 17 directing Southern Command to assert freedom of movement rights.
3 October	Panama Defense Force (PDF) dissidents stage an abortive coup against Noriega.
October-December	U.S. forces update plans for military action against Noriega and the PDF.
16 December	1stLt Robert Paz, USMC, killed by the PDF at a roadblock in Panama City.
20 December	D-Day for Operation Just Cause; Marine Forces Panama secures the western approaches to Panama City.
24 December	Noriega seeks refuge in the Papal Nunciatura in Panama City.

1990

3 January	Noriega surrenders to U.S. forces
January-April	U.S. forces participate in nation-building under Operation Promote Liberty.
13 June	Except for MCSF Marines, Marine Forces Panama embarks on board shipping.
23 June	Marine Forces Panama dissolved.

Appendix D
Unit Commendation

The Secretary of the Navy takes pleasure in presenting the NAVY UNIT COMMENDATION to
Marine Forces Panama
for service as set forth in the following

CITATION:

For exceptionally meritorious service during the volatile political crisis and combat operations in Panama from 1 April 1988 to 13 June 1990. Rapidly deployed to Panama during the attempted coup against the Noriega regime, Marine Forces Panama immediately found itself repelling armed intrusions into the Arraijan Tank Farm, several of which resulted in vicious exchanges of small arms fire. As the political situation worsened, the Marines exhibited superb discipline and courage during nearly 50 freedom of movement operations. These critical operations often confronted armed civil and Panamanian Defense Force opposition, but were essential to the protection of American lives and property. During Operation JUST CAUSE, Marine Forces Panama conducted combat operations against vastly superior forces over extended areas to secure the western approaches to Howard Air Force Base and key Panama Canal Defense Sites. In the ensuing months, Marine Forces Panama ranged throughout Panama, restoring law and order and rebuilding Panama's shattered countryside. By their inspiring courage, superior resourcefulness, and loyal devotion to duty, the officers and enlisted personnel of Marine Forces Panama reflected great credit upon themselves and upheld the highest traditions of the Marine Corps and the United States Naval Service.

/s/ H. Lawrence Garrett, III
Secretary of the Navy

Index

ISBN 0-16-048729-3
90000
9 780160 487293